Computational Aspects of Discrete Subgroups of Lie Groups

CONTEMPORARY MATHEMATICS

783

Computational Aspects of Discrete Subgroups of Lie Groups

Virtual Conference
Computational Aspects of Discrete Subgroups of Lie Groups
June 14–18, 2021
Institute for Computational and Experimental Research in Mathematics (ICERM)
Providence, Rhode Island

Alla Detinko
Michael Kapovich
Alex Kontorovich
Peter Sarnak
Richard Schwartz
Editors

AMS AMERICAN MATHEMATICAL SOCIETY

2020 *Mathematics Subject Classification.* Primary 20-04, 20-08, 20F67, 20G15, 20H25, 22E40, 53C35, 68W30.

Library of Congress Cataloging-in-Publication Data

Names: Detinko, Alla, editor.

Title: Computational aspects of discrete subgroups of lie groups : Virtual Conference on Computational Aspects of Discrete Subgroups of Lie Groups, June 14–18, 2021, Institute for Computational and Experimental Research in Mathematics (ICERM), Providence, Rhode Island / Alla Detinko, Michael Kapovich, Alex Kontorovich, Peter Sarnak, Richard Schwartz, editors.

Description: Providence, Rhode Island : American Mathematical Society, [2023] | Series: Contemporary mathematics, 0271-4132 ; volume 783 | Includes bibliographical references.

Identifiers: LCCN 2022042151 | ISBN 9781470468040 (paperback) | ISBN 9781470472610 (ebook)

Subjects: LCSH: Lie groups–Congresses. | Group theory–Congresses. | Computer algorithms–Congresses. | AMS: Group theory and generalizations – Explicit machine computation and programs. | Group theory and generalizations – Special aspects of infinite or finite groups – Hyperbolic groups and nonpositively curved groups. | Group theory and generalizations – Linear algebraic groups and related topics – Linear algebraic groups over arbitrary fields. | Group theory and generalizations – Other groups of matrices – Other matrix groups over rings. | Topological groups, Lie groups – Lie groups – Discrete subgroups of Lie groups. | Differential geometry – Global differential geometry – Symmetric spaces. | Computer science – Algorithms – Symbolic computation and algebraic computation.

Classification: LCC QA387 .C66 2023 | DDC 512/.55–dc23/eng20230111

LC record available at https://lccn.loc.gov/2022042151

Contemporary Mathematics ISSN: 0271-4132 (print); ISSN: 1098-3627 (online)

DOI: https://doi.org/10.1090/conm/783

---

10 9 8 7 6 5 4 3 2 1 28 27 26 25 24 23

Contents

Preface

This volume contains the proceedings of the workshop "Computational Aspects of Discrete Subgroups of Lie Groups", held at the Institute for Computational and Experimental Research in Mathematics (ICERM), June 14–18, 2021. The workshop's main theme interfaces algebra, geometry and computer science; it deals with the design, implementation, and application of algorithms based on matrix representations of groups and their geometric properties.

Lie groups and their subgroups have a long history, going back to the late 19th century. They impact many areas of mathematics, such as differential geometry, topology, number theory, algebraic geometry, differential equations, and combinatorics. The setting of *linear* Lie groups is suited to calculation and modeling transformations. At the same time, rapid technological development has realized the efficiency of matrix representation of groups and related algebraic structures in computers, leading to new ways of solving disparate problems via group-theoretical computer modeling.

The traditional theory of discrete subgroups of Lie groups (apart from Kleinian groups, discrete isometry groups of hyperbolic spaces) focused on *lattices*, i.e., discrete subgroups of finite covolume. Recently there has been significant progress in our understanding of "thin" discrete subgroups of matrix groups, which are discrete matrix groups that have infinite covolume in their Zariski closure. This was spurred by developments in several fields including geometry, dynamics, and number theory. The open problems encompass purely mathematical questions and those of more applied nature, relating to theoretical physics, quantum computing, and materials science. Many of the problems in the theory of discrete subgroups of Lie groups are, in principle, amenable to a computational approach.

While computer algebra systems provide a general computational framework, solution of particular problems requires building up foundations via new algorithms and software. This, in turn, necessitates collaboration between eclectic groups of scientists. Consequently a major goal of the workshop was to synergize and synthesize these independent strands. We aimed to facilitate solution of theoretical problems by means of recent advances in computational algebra, and stimulate development of computational algebra oriented to other mathematical disciplines.

The workshop featured four mini-courses and separate invited talks on major developments reflecting the workshop's theme. The lectures were supplemented by software demonstrations and tutorials covering relevant MAGMA, GAP, and SnapPea packages. The program incorporated problem sessions, and multidisciplinary discussions on development of computational algebra and computer algebra systems. Further information, including videos of talks, is available at the workshop website: `https://icerm.brown.edu/topical_workshops/tw-21-cads/`.

We thank everyone who participated, especially the contributors and referees. We are grateful to ICERM for its generous support. Also, without the hard work of the editorial staff of the American Mathematical Society, this volume would not have seen print. Our thanks go to Christine M. Thivierge for her constant help.

The Editors

Contemporary Mathematics
Volume **783**, 2023
https://doi.org/10.1090/conm/783/15744

Enumerating Kleinian Groups

David Gabai, Robert Meyerhoff, Nathaniel Thurston, and Andrew Yarmola

ABSTRACT. In this manuscript, we give an overview of the tools and techniques needed for successfully classifying "low-complexity" Kleinian groups. In particular, we focus on extracting topological and geometric properties of discrete Kleinian groups, such as bounds on tube radii, cusp geometry, volume, relators in group presentation, and similar quantities. A key point of this manuscript is to explain how a discrete set of solutions (or their closure) can be found using continuous methods, in particular by searching over a continuous parameter space of groups. These methods provide an effective avenue for studying and classifying hyperbolic 3-manifolds that satisfy some geometric or topological constraints.

1. Introduction

The geometrization theorem [**36**], [**37**] shows that closed oriented irreducible 3-manifolds decompose into pieces that each admit a geometric structure arising from one of eight Lie groups. Looking at the holonomy of these structures, one obtains discrete, cofinite subgroups of the corresponding Lie group. Of the eight types, Kleinian groups – corresponding to hyperbolic structures – have garnered a great deal of interest. In the past several decades, the study of Kleinian groups and, more generally, low-dimensional topology have also benefited from computational methods that provide both examples and computationally-assisted verified proofs. Our main goal with this manuscript is to illustrate an avenue, via examples and methods, for finding geometric constraints on discrete groups and extracting topological, algebraic, and geometric information that may be used to prove further results. As detailed in Section 2, these methods have had a number of notable successes and we believe that they have great potential for future development. Our examples here will be limited to Kleinian groups, that is discrete subgroups of

2020 *Mathematics Subject Classification.* Primary 57M50, 51M10, 51M25.

The first author was partially supported by NSF grants DMS-1006553, DMS-1607374 and DMS-2003892.

This work was partially supported by a grant from the Simons Foundation (#228084 to Robert Meyerhoff) and by a grant from the National Science Foundation (DMS-1308642 to Robert Meyerhoff).

The forth author was partially supported as a Visiting Student Research Collaborator with DMS-1006553 and Postdoctoral Researcher with DMS-1607374.

The authors also acknowledge the support and resources of the Polar Computing Cluster at the Mathematics Department and the Program in Applied and Computational Mathematics at Princeton University. The authors also thank David Futer for comments and corrections.

PSL(2, $\mathbb{C}$), as studied in [**20, 22**], but we expect these methods to have broader application, especially in low rank where Margulis Super-Rigidity and Arithmeticity do not apply.

Let G be a Lie group and $\Gamma \leq G$ a discrete group. One basic approach to study Γ is to focus on finitely generated subgroups that have specific properties. For example, if one is interested in finding maximal tubes about a short geodesic in a hyperbolic manifold, the subgroup of interest would be generated by an element representing a shortest geodesic and a conjugating element that corresponds to a self-tangency of the tube. A similar approach can be taken for maximal cusps, or for analyzing Margulis numbers. Most of our examples focus on 2 or 3 generator subgroups as their parameter spaces tend to be small enough for computational tools to work, yet too large for by-hand analysis. The goal of the methods we describe here is to classify which part of the parameter space contains all discrete groups and to extract useful information about these groups. This information includes provable relators in the group, bounds on the generators, and other geometric and algebraic output. In Section 2, we give examples of results that can be directly extracted for such searches and some relevant historical background.

The basic idea is to eliminate places in the parameter space where some geometric measurement violates discreteness or other assumptions. Starting with Section 4, the text is organized as follows. We first introduce the notion of *marking* and give examples of how to choose one. Next, we discuss how to set up the parameter space itself. The hard work begins with trying to understand how to cut away "bad" portions of the parameter space, that is, remove places where discrete groups cannot exist (or some other assumptions about the marking fail). Here, words in the generators play a key role. Outside of helping one eliminate parameters, words can be used to find guaranteed relators for the discrete groups that appear in the parameter space. We explain this dichotomy when discussing *killer*, *quasi-relator* and *variety* words. In Section 7, we get down to the details of how to computationally partition the parameter space into boxes in an organized way as to make sure that the problem of finding the aforementioned words has a chance of being feasible. In Section 8, we discuss how to build word-finding algorithms and their pitfalls. Finally, in Section 9, we give an overview of our preferred choice of arithmetic and in Section 10, we discuss methods for sanity checking progress.

REMARKS 1.1. It is worth pointing out that most of this paper deals with "large" parameter spaces. In particular, we aren't concerned with studying a specific manifold or group most of the time. However, it is often possible to get information for specific manifolds as a result. For example, we can use relators found for the Whitehead link to show that the maximal cusp volume for that specific manifold decreases under Dehn filling of one of the cusps. Similarly, it should be possible to compute more specific rigorous information for hyperbolic manifolds, such as length spectra, upper and lower bounds on volume, and canonical triangulations.

It is important to note that for studying specific manifolds, results along these lines are spectacularly provided by SnapPea [**42**] and its descendants such as Snap [**14, 25**] and SnapPy [**16**]. In its current incarnation, SnapPy is indispensable to workers in the field and can produce verified results for several computations, such as volume and isometry validation. Another very useful tool in this area is Regina [**12**], great for working with triangulations and normal surfaces.

2. Applications and historical context

In the context of closed hyperbolic 3-manifolds, the existence of sufficiently thick tubes about short geodesics is crucial to many results in the subject. One such result is the $\log(3)/2$-*theorem* of [**22**], which was proven using techniques discussed in this manuscript. The sharp form of that result [**24**] asserts that with the exception of the manifold known as Vol3, any closed hyperbolic 3-manifold X has a $\log(3)/2$ tube about some geodesic; furthermore, with five exceptions this geodesic can be taken to be any shortest one. Results that crucially use this work include the Smale conjecture for hyperbolic 3-manifolds [**19**] and the important inequality (slightly updated to reflect [**24**]), that if $X \neq$ Vol3 is a closed hyperbolic 3-manifold, then it is obtained by filling a 1-cusped hyperbolic 3-manifold Y such that $\mathrm{vol}(Y) < 3.0177\,\mathrm{vol}(X)$ [**6**]. This inequality is based on earlier works of Agol [**4**] and Agol-Dunfield [**7**] and uses Perelman's work on Ricci flow [**36**], [**37**]. It is needed to show that the Weeks manifold is the unique closed hyperbolic 3-manifold of least volume [**23**].

Each end of a complete, orientable, finite-volume hyperbolic 3-manifold Y is homeomorphic to $T^2 \times (1, \infty)$. These ends are called *cusps*. Each cusp contains a properly embedded *horocusp*, which is a region isometric to a quotient of H_∞, where $H_\infty = \{(x, y, t) : t > 1\} \subset \mathbb{H}^3$ in the upper half-space model and we quotient by a group of translations of (x, y) isomorphic to $\mathbb{Z} \oplus \mathbb{Z}$, see [**40**]. Further, this region can be maximally enlarged in the manifold to a *maximal horocusp* κ for that end. Here, κ is embedded in Y and $\partial\kappa$ has finitely many self-tangencies. A basic fact is that $\mathrm{vol}(\kappa) = \mathrm{area}(\partial\kappa)/2$, called the *cusp volume*. Note, the term cusp in this paper always refers to a *rank-two* cusp, even if the manifold in question is not finite-volume.

The works of Jørgensen and Thurston from the 1970's demonstrate the close relation between thick tubes about geodesics and maximal horocusps of complete manifolds [**39**]. Indeed, Jørgensen showed that after passing to a subsequence, any infinite sequence Y_i of distinct complete hyperbolic 3-manifolds of uniformly bounded volume limits geometrically to a complete manifold Y_∞, where thicker and thicker tubes of the Y_i's limit to cusps of Y_∞. Conversely, Thurston showed that given a cusp of a complete finite-volume 3-manifold Y and $\epsilon > 0$, then all but finitely many fillings on that cusp produce hyperbolic manifolds geometrically ϵ-close to Y. The papers, [**35**], [**29**], [**30**],[**18**] have made this connection more explicit and quantitative.

Let κ be a maximal horocusp of the complete hyperbolic 3-manifold $Y = \mathbb{H}^3/\Gamma$. Up to conjugating Γ, we assume that H_∞ and H_0 both cover κ and are conjugate under Γ, where H_0 is a horoball centered at 0 and tangent to H_∞. Let $\mathfrak{B} = \langle m, n, g\rangle \leq \Gamma$, where m, n are generators of $\{\gamma \in \Gamma : \gamma \cdot H_\infty = H_\infty\}$ and g is a conjugating element $g(H_\infty) = H_0$. By studying the parameter space of such 3-generator groups, the authors in [**20**] prove the following two results.

THEOREM 2.1. *The figure-8 knot complement and its sister are the complete hyperbolic 3-manifolds with a maximal horocusp of minimal volume. This volume is* $\sqrt{3}$.

REMARKS 2.1.

(i) These manifolds minimize cusp volume among all complete hyperbolic 3-manifold with torus cusps, even among multi-cusped manifolds and complete manifolds of infinite volume.
(ii) Results on this question go back to the 1980's. A lower bound of $\sqrt{3}/4$ for the volume of a maximal cusp was obtained in [**33**]. Adams improved this bound by a factor of 2 in [**2**]. There were no further improvements until Cao and Meyerhoff [**13**] gave a bound of 3.35/2 for the volume of a maximal cusp in a hyperbolic 3-manifold, which is within 0.0571 of the actual value of $\sqrt{3}$.

THEOREM 2.2. *If Y is a complete hyperbolic 3-manifold with a maximal horocusp of volume ≤ 2.62, then $\pi_1(Y) = \mathfrak{B} = \langle m, n, g\rangle$ and $\text{vol}(Y) < \infty$. Furthermore, there exists a reduced relation $w(m, n, g)$ where g and g^{-1} appear at most 7 and at least 4 times in total.*

This gives an effective analogue of a theorem of Agol [**5**], which shows that some relator always exists whenever $\text{vol}(\kappa) < \pi$.

By applying Theorem 2.2, the authors of [**20**] are able to give a finite list of manifolds whose Dehn fillings give all manifolds with maximal cusp volume at most 2.62. This also shows that all such manifolds are tunnel number one. Enumerating small volume manifolds is also possible using this result. By a packing argument (see [**34**]), if Y has a maximal horocusp of volume V, then $\text{vol}(Y) \geq (2v_3/\sqrt{3})V \approx (1.17195\ldots)V$, where v_3 is the volume of the regular ideal tetrahedron in $\mathbb{H}^3$. It follows that either $\text{vol}(Y) \geq 2.62 \cdot (1.17195\ldots) \approx 3.0705\ldots$ or Y is obtained by filling one of the manifolds listed in [**20**]. This further allows for the classification the three smallest volume hyperbolic 3-manifolds.

In his revolutionary work in the 1970's on the geometry of the figure-8 knot complement, Thurston showed that it has exactly 10 non-hyperbolic Dehn fillings. Thurston's result spawned a tremendous amount of work towards understanding hyperbolic and non-hyperbolic fillings of hyperbolic 3-manifolds, e.g. see the surveys in [**26**], [**11**], [**31**]. In particular, there are infinitely many 1-cusped hyperbolic manifolds with six non-hyperbolic fillings. By 1998, there were (resp. eight, two, zero, one) known 1-cusped manifolds with (resp. seven, eight, nine, ten) exceptional surgeries. This led Gordon to conjecture in [**26**] that if Y is a hyperbolic 3-manifold with boundary a torus, then Y has at most 8 non-hyperbolic fillings unless Y is the figure eight knot exterior.

Let $n(Y)$ denote the number of non-hyperbolic fillings of a 1-cusped hyperbolic manifold Y. The Gromov-Thurston 2π-theorem showed that if Y' is obtained by filling Y along a curve $\gamma \subset \partial\kappa$ with the length $\ell_\kappa(\gamma)$ on $\partial\kappa$ greater than 2π, then Y' has a complete metric of negative sectional curvature (hence a hyperbolic structure by Perelman). Using the fact that $\ell_\kappa(\gamma) \geq 1$ for all closed curves on $\partial\kappa$, Thurston showed that $n(Y) \leq 48$ for all Y. However, coupling a lower bound on $\ell_\kappa(\gamma)$ with a lower bound on $\text{vol}(\kappa)$ can lead to an improvement. This was carried out by Bleiler and Hodgson [**10**]—using Adams's bound of $\text{vol}(\kappa) \geq \sqrt{3}/2$—to obtain that $n(Y) \leq 24$. Had the Cao-Meyerhoff bound of $\text{vol}(\kappa) \geq 3.35/2$ been available at that time, then these techniques give $n(Y) \leq 14$ for all Y.

In 2000, Agol and Lackenby independently improved the 2π-theorem as follows: if $\ell_\kappa(\gamma) > 6$, then one obtains the weaker conclusion that Y' has non-elementary and word-hyperbolic fundamental group (again Perelman implies that Y' is hyperbolic). Coupling the 6-theorem with the Cao-Meyerhoff bound of $\mathrm{vol}(\kappa) \geq 3.35/2$ led to $n(Y) \leq 12$—tantalizingly close to the figure-eight case bound of 10. As the 6-theorem is sharp, attention was focused on maximal cusp volume bound improvements. However, a new bound of $\mathrm{vol}(\kappa) \geq 3.7/2$ in the "non-Mom" case in [**23**] did not improve upon the non-hyperbolic filling bound of 12. This new lower bound on $\mathrm{vol}(\kappa)$ seemed quite strong at the time, so potential for progress on the non-hyperbolic filling bound seemed bleak. However, by generalizing the 6-theorem and exploiting Mom technology directly, Lackenby and Meyerhoff [**31**] were able to prove that $n(Y) \leq 10$ for all Y.

To finally answer Gordon's conjecture, a few more results were required. In [**3**] and [**5**], Agol proved that if Y is a 1-cusped hyperbolic 3-manifold $n(Y) \geq 9$, then $\mathrm{vol}(\kappa) > 18/7 = 2.57\ldots$. Since $18/7 < 2.62$, all the 1-cusped hyperbolic 3-manifolds with a maximal cusp of volume $\leq 18/7$ are Dehn filling of those given in [**20**], which after a Dehn filling analysis, also given by Crawford [**15**], proves Gordon's conjecture.

THEOREM 2.3. *Let M be a hyperbolic 3-manifold with boundary a torus. Then, M has at most 8 non-hyperbolic fillings unless M is the figure eight knot exterior.*

3. Background

Our examples will focus on hyperbolic 3-manifolds. We will use the upper-half-space model $\mathbb{H}^3 = \{(z,t) \in \mathbb{C} \times \mathbb{R} : t > 0\}$ of hyperbolic 3-space. Consider $q \in \mathbb{H}^3$ in quaternion notation $q = x + \mathfrak{i}y + \mathfrak{j}t$, where $\mathfrak{i}^2 = \mathfrak{j}^2 = \mathfrak{k}^2 = \mathfrak{i}\mathfrak{j}\mathfrak{k} = -1$. Identifying $\mathrm{Isom}^+(\mathbb{H}^3) \cong \mathrm{PSL}(2,\mathbb{C})$, the action on $q = x + \mathfrak{i}y + \mathfrak{j}t$ is given by

$$\pm\begin{pmatrix} a & b \\ c & d \end{pmatrix} \cdot q = (aq+b)(cq+d)^{-1}.$$

Notice that for $z \in \mathbb{C}$, we have that $z\,\mathfrak{j} = \mathfrak{j}\,\overline{z}$, so one computes

$$\pm\begin{pmatrix} a & b \\ c & d \end{pmatrix} \cdot t\,\mathfrak{j} = (a\,t\,\mathfrak{j} + b)(c\,t\,\mathfrak{j} + d)^{-1} = \frac{t^2\,a\,\overline{c} + b\,\overline{d}}{t^2|c|^2 + |d|^2} + \frac{t\,\mathfrak{j}}{t^2|c|^2 + |d|^2}. \tag{1}$$

In particular, $\mathfrak{j}$ is mapped to a point of Euclidean height $1/(|c|^2 + |d|^2)$, which we will use later.

The group $\mathrm{Isom}^+(\mathbb{H}^3)$ acts transitively on $\mathbb{H}^3$ with point stabilizer $\mathrm{SO}(3)$. On the boundary at infinity, thought of as the Riemann sphere $\hat{\mathbb{C}} = \partial_\infty \mathbb{H}^3$, the action is transitive on triples of points. Elements of $\mathrm{Isom}^+(\mathbb{H}^3)$ fall into three classes based on their fixed points in $\overline{\mathbb{H}}^3$: *loxodromic* elements fix two points on $\hat{\mathbb{C}}$, *parabolic* elements fix one point on $\hat{\mathbb{C}}$, and *elliptic* elements fix points on the interior $\mathbb{H}^3$. Discrete groups $\Gamma \leq \mathrm{Isom}^+(\mathbb{H}^3) \cong \mathrm{PSL}(2,\mathbb{C})$ are called *Kleinian groups*. If they are torsion-free, then they contain no elliptics. Any subgroup of Γ made up of only parabolics has rank at most 2. Further, if $\mathbb{H}^3/\Gamma$ is compact, then Γ contains no parabolics. The action of a loxodromic preserves a complete geodesic in $\mathbb{H}^3$, called its axis. This action about the axis is a composition of a non-trivial translation and a rotation, giving loxodromics the notion of complex lengths, where the real part is the translation and the imaginary is the rotation. For a parabolic, if one conjugates its fixed point to $\infty \in \hat{\mathbb{C}}$, its action on $\hat{\mathbb{C}}$ looks like $z \mapsto z + c$ for $c \in \mathbb{C}$. On the interior, this action preserves *horoballs at infinity*, which are sets of the form

$\mathbb{H}_\infty = \{(z,t) : t > a\}$ for some real $a \in \mathbb{R}^+$. The image of H_∞ under an element not fixing ∞ looks like a Euclidean ball tangent to $\mathbb{C}$ at a point, called its *center*.

Geodesics of $Y = \mathbb{H}^3/\Gamma$ correspond to conjugacy classes in Γ, and *tubes* about geodesics in Y corresponds to regular neighborhoods of axes of loxodromics in $\mathbb{H}^3$, also often called tubes. A maximal tube of Y lifts to a collection of tangent tubes upstairs. We often only focus on one tangency at a time and this corresponds to an element $w \in \pi_1(Y) = \Gamma$ that maps one tube in a tangent pair to the other. When Y has finite volume, its ends are of the form $T^2 \times (1, \infty)$ and contain embedded neighborhoods isometric to H_∞/Π, where Π is a group of parabolics fixing infinity and isomorphic to $\mathbb{Z} \oplus \mathbb{Z}$. Such neighborhoods are called *horocusps*. Just like for tubes, tangencies of maximal horocusps correspond to elements of $\pi_1(Y)$.

4. Markings

In this section, we introduce the notion of a *marking*, which plays a central role in our techniques for analyzing parameter spaces of subgroups of $\mathrm{PSL}(2,\mathbb{C})$. We will focus on the mathematical setup and examples before discussing the computational aspects in later sections. The general idea is to look at groups generated by a *marked* set of generators. Such markings, defined below, will then allow us to (1) parametrize all possible generating sets and (2) efficiently eliminate ranges of generating sets from consideration. This will leave us with some (possibly large) collection of pieces of the original parameter space that includes all (correctly marked) discrete groups of interest, allowing us to finally extract topological, geometric, combinatorial and algebraic information.

4.1. Markings and examples.

4.1. Markings and examples. Fix a Lie group G. For us, this will almost always be $\mathrm{PSL}(2,\mathbb{C})$.

DEFINITION 4.1. A *marking* in a Lie group G is the tuple $(S, U, \mathfrak{Dicta})$, where $S = \{g_1, \ldots, g_k\} \subset G$ is an ordered set, U is a subset of a homogenous space of G, and $\mathfrak{Dicta}$ is a collection of desirable properties expressed in terms of S and U. We will let Γ_S denote the (possibly indiscrete) group $\langle g_1, \ldots, g_l \rangle$. $\mathfrak{Dicta}$ should be chosen such that if all properties hold for Γ_S, then it is a group of interest. Further, if some property in $\mathfrak{Dicta}$ *fails* for a particular Γ_S, it should be possible to verify this in finite time.

Examples. While the definition of marking is rather informal, we hope the following examples will make this notion relatively clear. Specifically, $\mathfrak{Dicta}$ must be carefully designed to work in the relevant context and be precise enough to pinpoint the groups of interest.

Tubes. Suppose we would like to find the largest radius of a tube that is guaranteed to be embedded around a shortest geodesic in a closed hyperbolic 3-manifold Y. Let γ be a shortest geodesic in Y and consider a maximal embedded (open) tube R around γ, so by maximality its boundary has a self-tangency. In the universal cover $\mathbb{H}^3$ of Y, this picture corresponds to a tube U around a complete geodesic (i.e lifts of R and γ, respectively) and a tube tangent to U of the form $w \cdot U$ for some $w \in \pi_1(Y)$. From this, we build our marking as follows. Take $G = \mathrm{PSL}(2,\mathbb{C})$, U a tube stabilized by a loxodromic f with $U/\langle f \rangle = R$,and $S = \{f, w\}$.

We will want to think of f as corresponding to our shortest geodesic γ in Y. So, we take $\mathfrak{Dicta}$ to be the following set of assumptions:

(1) f, w are non-commuting loxodromics,
(2) the core geodesic of U is the axis of f,
(3) U and $w \cdot U$ are tangent,
(4) f has smallest real translation length amongst $\Gamma_S \setminus \{id\}$,
(5) Γ_S is torsion-free, and
(6) for any $h \in \Gamma_S \setminus \langle f \rangle$, the interiors of U and $h \cdot U$ are disjoint.

Conditions (5) and (6) imply that Γ_S is a torsion-free Kleinian group and $R = U/\langle f \rangle$ is our maximal embedded tube around γ in $Y = \mathbb{H}^3/\Gamma_S$. Note, conditions such as (1)-(3) can often be encoded in the choice of f, w and U when we consider setting up a parameter space of markings. This marking was used in [**22**] to prove that outside of a few exceptions, the tube radius can be taken to be at least $\log(3)/2$.

Cusps. Similarly, suppose you want to study maximal cusp neighborhoods and their geometry in a finite-volume hyperbolic 3-manifold Y. Any such maximal cusp neighborhood has a horoball lift in $\mathbb{H}^3$. This lift is then preserved by some pair of parabolics in $\pi_1(Y)$ and, by maximality, is moved to a tangent horoball by some element in $\pi_1(Y)$. Taking this into account, we let $S = \{m, n, g\}$ and U a horoball in $\mathbb{H}^3$. Then $\mathfrak{Dicta}$ are the conditions:

(1) m, n are commuting non-parallel parabolics,
(2) m, n preserve U,
(3) m has shortest translation length on ∂U amongst $\langle m, n \rangle \setminus \{id\}$,
(4) $g \cdot U$ is tangent to U,
(5) Γ_S is torsion-free, and
(6) the interiors $w \cdot U$ and U are disjoint for all $w \in \Gamma_S \setminus \langle m, n \rangle$.

Notice that if any of these conditions fail, they can be computationally verified to fail. This marking was used in [**20**] to classify the infinite family of cusped manifolds with maximal cusp volume less than or equal to 2.62 and to show that the figure eight knot and its sister minimize maximal cusp volume. Forthcoming work [**21**] will give improved bounds on the distance between exceptional slopes in 1-cusped hyperbolic manifolds.

Margulis. Recall that the Margulis number of a Kleinian group Γ is the number

$$\begin{aligned}\mu_\Gamma = \sup\{d \in \mathbb{R}^+ : \\ \text{if } d_{\mathbb{H}^3}(p, x_i \cdot p) < d \text{ for } x_1, x_2 \in \Gamma \text{ and } p \in \mathbb{H}^3, \text{ then } x_1, x_2 \text{ commute}\}\end{aligned}$$

Say we want to find the smallest Margulis number amongst hyperbolic 3-manifolds, called the Margulis constant. To accomplish this, we take $S = \{x_1, x_2\}$, $U = p$ a point, and $\mathfrak{Dicta}$ is the set of conditions:

(1) x_1, x_2 do not commute,
(2) if $d_{\mathbb{H}^3}(p, w \cdot p) < \max_i d_{\mathbb{H}^3}(p, x_i \cdot p)$ for some $w \in \Gamma_S$, then w commutes with x_1 or x_2,
(3) $\max_i d_{\mathbb{H}^3}(q, w_i \cdot q) \leq \max_i d_{\mathbb{H}^3}(p, x_i \cdot p)$ for any non-commuting pair $w_1, w_2 \in \Gamma_S$ and $q \in \mathbb{H}^3$,
(4) Γ_S is torsion-free, and
(5) x_1 has real translation length at most that of x_2.

When Γ_S is discrete, (1) and (2) impliy that $\mu_{\Gamma_S} \leq \max_i d_{\mathbb{H}^3}(p, x_i{\cdot}p)$. Together with (3), we get that $\mu_{\Gamma_S} = \max_i d_{\mathbb{H}^3}(p, x_i{\cdot}p)$, so x_1, x_2 and p realize the Margulis number of Γ_S. It is interesting to note that (3) may be dropped depending on the context. If the goal is to find the Margulis constant, removing (3) would result in a larger family of markings, but any marking where μ_{Γ_S} is the Margulis constant would still be present. We also note that (5) allows us to reduce the size of the parameter space of markings. Work on this parameter space with the goal of showing that the Weeks manifolds uniquely realizes the Margulis constant is currently in progress [**17**].

Sys. For hyperbolic surfaces, one may want to find the maximal injectivity radius of a surface with fixed topology. Here, one would take U to be empty and construct S in some systematic way from parameters on moduli space (e.g. Fenchel-Nielsen coordinates for some fixed pants decomposition). Then $\mathfrak{Dicta}$ is the condition that the injectivity radius of Γ_S is at least some known number. As one finds coordinates for which Γ_S has larger and larger injectivity radii, the space of possible S can be further restricted until it is small enough to analyze by hand. As we shall see later, our approach and proof technique does indeed allow for such iterative methods, though this particular one is in early stages of implementation [**41**] unlike previous examples.

4.2. Geometric and incorrect markings. We call a marking *geometric* if it arises from a hyperbolic manifold (or a discrete group of interest). That is, if Γ_S is discrete, torsion-free, and $\mathfrak{Dicta}$ is satisfied for the given ordered generating set S and the set U. Our objective is to find bounds on S and U that come from geometric markings, or, even better, prove that geometric markings of a particular type must satisfy some collection of relators between elements of S. For example, in [**20**], it is show that any geometric marking in Cusps with cusp-volume at most 2.62 must have a (cyclically reduced) relator between m, n, g where the total occurrence of g or g^{-1} is at most 7.

Remark 4.2. The requirement of torsion-free appears here only because all of our examples are focused on hyperbolic 3-manifolds. One could just as easily drop this condition to look at orbifolds. In fact, it would be interesting to see if Marshall and Martin's proof that finds the two smallest volume hyperbolic-3 orbifolds [**32**] could be re-proven using techniques described here.

We now address the utility of markings. Consider Example Tubes, which looks for embedded tubes about shortest geodesics. Assume we start with some given $f, w \in \mathrm{PSL}(2, \mathbb{C})$ and tube U around the axis of f. We can then begin to search for possible contradictions to $\mathfrak{Dicta}$. In particular, if we find an element $g \in \Gamma_S \setminus \{id\}$ whose real translation length is provably less than that of f, then we know that the marking we started with cannot be geometric, and therefore can be ignored. Similarly, if we find an element $h \in \Gamma_S \setminus \langle f \rangle$ such that the interiors of U and $h \cdot U$ provably intersect, then we either violate discreteness, the fact that f must be primitive to be shortest, or the embeddedness of the tube corresponding to U, so again, the marking is not geometric. Both types of these contradictions are a bit subtle. It is possible that Γ_S is, in fact, discrete but simply our choice of f or U and w was incorrect. If we take care in building a large enough parameter space of markings, then we can assume a correct marking for this discrete group Γ_S lies *elsewhere* in the parameter space. In particular, this marking can be ignored.

DEFINITION 4.3. A marking is *incorrect* if some condition in $\mathfrak{Dicta}$ is not satisfied by S and U.

Another illustration of incorrect markings can be found in Example MARGULIS. Say that we find a word $w \in \Gamma_S \setminus \{id\}$ that moves our point $U = p$ less that x_1 or x_2, then even if Γ_S is discrete, this means that we chose the wrong pair x_1, x_2 to realize the Margulis number of Γ_S. Similarly, if we find a non-commuting pair $w_1, w_2 \in \Gamma_S$ and a point q for which w_1 and w_2 move q less than x_1 and x_2 move p, then our marking is incorrect and can be ignored. In particular, if Γ_S is discrete and w_1, w_2 actually realize its Margulis number, then a correct marking has $S = \{w_1, w_2\}$ and $U = q$. As remarked previously, in the actual search, we do not necessarily need to look for such w_1, w_2, q for the purposes of eliminating parameters if our goal is just to find the Margulis constant.

Before we construct parameters for markings, let us define the notion of equivalence for markings.

DEFINITION 4.4. Two markings $(S_1, U_1, \mathfrak{Dicta}_1)$ and $(S_2, U_2, \mathfrak{Dicta}_2)$ in G are said to be equivalent if $\Gamma_{S_2} = \mu\Gamma_{S_1}\mu^{-1}$ for some $\mu \in G$, $U_2 = \mu \cdot U_1$, and, with $S_1 = \{g_1, \dots, g_k\}$ and $S_2 = \{h_1, \dots, h_k\}$, we have that $\mathfrak{Dicta}_1 = \mathfrak{Dicta}_2$ under the replacement of g_i by h_i for $i = 1, \dots, k$ and U_1 by U_2.

Basic examples of equivalence are conjugation in $\mathrm{PSL}(2, \mathbb{C})$, so that $\mu g_i \mu^{-1} = h_i$, or replacing a generator in S by its inverse (as long as that doesn't break the validity of $\mathfrak{Dicta}$). Notice that in the latter case, $\mu = id$. In fact, $\mu = id$ is quite common and is used when replacing generators with better ones. The discussion in the next section shows how we can use the notion of equivalence to normalize our markings so that the space of parameters can be made small.

5. Parameter spaces

The next step is to build a parameter space $\mathcal{P} \subset \mathbb{C}^d$ that is large enough to include parameters representing each equivalence class of the geometric markings of interest. When picking $\mathcal{P}$, we want to make sure that we can recover S and U (up to equivalence) from $\mathfrak{p} \in \mathcal{P}$ and, for computational purposes, that each equivalence class of markings appears once or only a few times in $\mathcal{P}$.

For example, if we want to prove that the figure-eight knot complement and its sister are the two manifolds with smallest maximal cusp volume, then we need to consider markings of type CUSPS that have maximal cusp volume at most $\sqrt{3}$. Our tasks will be (1) to show that this parameter space can be chosen to be compact up to equivalence of markings, (2) computationally eliminate large regions of this parameter space from consideration, and (3) show that the remaining regions only contain markings corresponding to the figure-eight knot complement and its sister. Tasks (2) and (3) will be addressed in Section 6. The goal of this section is to explain (1) in detail as this is a prototype for addressing a wide range of geometric problems.

5.1. Parameters for Cusps. Here is how such a parameter space is constructed. Our generators are $\{m, n, g\}$ with m and n commuting non-parallel parabolics and U is a horoball centered at the fixed point of m and n. Under our $\mathfrak{Dicta}$ assumptions, we now do a sequence of transformations to find an equivalent marking that is nicely normalized. Working in the upper-half-space model of $\mathbb{H}^3$, we will conflate elements of $\mathrm{PSL}(2,\mathbb{C})$ with orientation preserving Möbius transformations and orientation-preserving isometries of $\mathbb{H}^3$. Note, there is some isometry ψ taking U to a horoball at infinity. Replacing m, n, g with their conjugates under ψ, we may assume that every element of $\langle m, n\rangle$ is of the form $z \mapsto z + c$ for some $c \in \mathbb{C}$. Let $|c|$ denote the length of such an element. To satisfy $\mathfrak{Dicta}$, we take $m(z) = z + \lambda$ to be a shortest nontrivial element of $\langle m, n\rangle$. Conjugating by $z \mapsto \lambda^{-1}z$, we may assume that m is the map $z \mapsto z + 1$. Further, we can take $n(z) = z + \ell$ to be a shortest element of $\langle m, n\rangle$ linearly independent of m (i.e. *next-shortest*). Notice that we did not add this condition on n to $\mathfrak{Dicta}$ for CUSPS. Since we are looking for representatives of equivalence classes of markings for our parameter space, we do not need to add such normalizations of generators to $\mathfrak{Dicta}$. As we shall see in the next section, the main use of the conditions in $\mathfrak{Dicta}$ is to eliminate parts of the parameter space. So, while the assumption that m is shortest will be used for elimination, the fact that n is next-shortest is only here to make the parameter space compact. Returning to our choice of generators, we have that $g \cdot U$ is a horoball tangent to U and centered at some point $c \in \mathbb{C}$. Conjugating by $z \mapsto z - c$, we may assume $g \cdot U$ is centered at 0. Note, once again, that this is just a choice for the parametrization and does not need to be part of $\mathfrak{Dicta}$. With these normalizations, we have $g^{-1}(z) = p + 1/(s^2 z)$ for some complex numbers p and s, where p is the center of the horoball $g^{-1}U$ and s controls its height and rotation.

To summarize, each equivalence class of markings of type CUSPS can be assigned a *parameter* $\mathfrak{p} = (p, s, \ell) \in \mathbb{C}^3$ with $s \neq 0$, were we can build a representative marking from this parameter by letting

$$g_{\mathfrak{p}}^{-1} = \pm\begin{pmatrix} ps\,\mathrm{i} & \mathrm{i}/s \\ s\,\mathrm{i} & 0 \end{pmatrix} \qquad g_{\mathfrak{p}} = \pm\begin{pmatrix} 0 & -\mathrm{i}/s \\ -s\,\mathrm{i} & ps\,\mathrm{i} \end{pmatrix}$$

$$m_{\mathfrak{p}} = \pm\begin{pmatrix} 1 & 1 \\ 0 & 1 \end{pmatrix} \qquad n_{\mathfrak{p}} = \pm\begin{pmatrix} 1 & \ell \\ 0 & 1 \end{pmatrix}$$

and choosing $U_{\mathfrak{p}}$ as follows. Since we want $g_{\mathfrak{p}} \cdot U_{\mathfrak{p}}$ to be tangent to $U_{\mathfrak{p}}$, by equation (1), we must have

$$U_{\mathfrak{p}} = \{(z,t) \in \mathbb{H}^3 : t > 1/|s|\}.$$

In our parametrization, when the marking is geometric, we have $\mathrm{vol}(U_{\mathfrak{p}}/\Gamma_{S_{\mathfrak{p}}}) = |s^2\,\mathrm{im}(\ell)|/2$. Observe that in inverting this construction, we have guaranteed that some but not all of the $\mathfrak{Dicta}$ assumptions are satisfied. In particular, we know that m, n are commuting non-parallel parabolics and g is a loxodromic with $g \cdot U$ tangent to U. In the next few sections, we will use the rest of the $\mathfrak{Dicta}$ conditions to eliminate nearly all sets of parameters that give rise to incorrect markings.

We can now show that given a geometric marking of the type CUSPS with the restriction that the maximal cusp volume is $\leq \sqrt{3}$, then it is equivalent to a marking whose parameter lies in the following compact set $\mathcal{P}_{\mathrm{CUSPS}} \subset \mathbb{C}^3$.

DEFINITION 5.1. Let $\mathcal{P}_{\mathrm{CUSPS}}$ be the subset of $\mathbb{C}^3$ defined by the following conditions:

(0) $|s| \geq 1$
(1) $\mathrm{im}(s) \geq 0$, $\mathrm{im}(\ell) \geq 0$, $\mathrm{im}(p) \geq 0$, $\mathrm{re}(p) \geq 0$
(2) $-1/2 \leq \mathrm{re}(\ell) \leq 1/2$
(3) $|\ell| \geq 1$
(4) $\mathrm{im}(p) \leq \mathrm{im}(\ell)/2$

(5) $\text{re}(p) \leq 1/2$
(6) $|s^2 \, \text{im}(\ell)| \leq 2\sqrt{3}$

PROPOSITION 5.2. *Suppose $(\{m, n, g\}, U, \mathfrak{Dicta})$ is a geometric marking of type* CUSPS *with a maximal cusp of volume at most $\sqrt{3}$. Then, there is $\mathfrak{p} = (p, s, \ell) \in \mathcal{P}_{\text{CUSPS}}$ such that $(\{m, n, g\}, U, \mathfrak{Dicta})$ is equivalent to $(\{m_{\mathfrak{p}}, n_{\mathfrak{p}}, g_{\mathfrak{p}}\}, U_{\mathfrak{p}}, \mathfrak{Dicta})$. Further, $\mathcal{P}_{\text{CUSPS}}$ is compact.*

PROOF. As above, we may assume U is centered at ∞ and $m(z) = z + 1$ is a shortest-length generator and $n(z) = z + \ell$ is the next-shortest in $\langle m, n \rangle$. Since n is next-shortest, we have $|\ell| \geq 1$ and $-1/2 \leq \text{re}(\ell) \leq 1/2$. If $\text{im}(\ell) < 0$, then we replace n with $z \mapsto z - \ell$, which is also next-shortest and has $\text{im}(-\ell) > 0$.

Next, let $p \in \mathbb{C}$ be the center at infinity of the horoball $g^{-1} \cdot U$. By post-composing g with elements of $\langle m, n \rangle$, we may choose g such that $-\text{im}(\ell)/2 \leq \text{im}(p) \leq \text{im}(\ell)/2$ and $-1/2 \leq \text{re}(\ell) \leq 1/2$. Reflecting across the x-axis creates isomorphic groups with isometric quotients. Thus we may assume $0 \leq \text{im}(\ell) \leq \text{im}(L)/2$. Reflecting across the y-axis allows us to assume likewise that $0 \leq \text{re}(\ell) \leq 1/2$. Finally, changing s to $-s$ leaves f invariant, so we may assume $\text{im}(s) \geq 0$. This determines p, s, and ℓ and a marking isomorphism. Properties (1), (2), (4), and (5) clearly hold and (3) holds since n is next-shortest. Further, since U/Γ_S is a maximal cusp neighborhood in $\mathbb{H}^3/\Gamma_S$, we know that U has Euclidean height ≤ 1. Thus, $1/|s| \leq 1$ and (0) holds.

Finally, the volume of U/Γ_S is $|s^2 \, \text{im}(\ell)|/2$, and this is at most $\sqrt{3}$, so (6) holds. Since all the inequalities are satisfied, $(p, s, \ell) \in \mathcal{P}$ as desired.

Note that the positive lower bounds for $|s|$ and $|\text{im}(\ell)|$ impose upper bounds for $|s|$ and $|\text{im}(\ell)|$ by using $|s^2 \, \text{im}(\ell)| \leq \sqrt{3}$. Thus $\mathcal{P}_{\text{CUSPS}}$ is compact. □

We hope that the above explicit example shows how to map markings to parameter spaces in general and what ideas to keep in mind. A good exercise is to work out a parameter space for TUBES. See **[22]** for details. Most importantly, we are now restricted to working with a bounded and explicitly described set of parameters $\mathcal{P}$. In the next section, we will discuss how to eliminate portions of this parameter space by looking for elements of Γ_S that violate the assumptions of $\mathfrak{Dicta}$.

REMARK 5.3. Our example looks at complex parameters because they are computationally more efficient in this setting. Using real parameters is possible, but requires computational considerations due to the choice of verified arithmetic used. This is one of several computational considerations that make the problems addressed in this paper computationally feasible. See Sections 7, 8 and 9 for more techniques.

6. Analyzing parameter spaces and elimination criteria

Naturally, we are interested in parameters $\mathfrak{p} \in \mathcal{P}$ where $\Gamma_{S_{\mathfrak{p}}}$ is geometric. To investigate these points it is convenient to introduce the following subsets of $\mathcal{P}$.

DEFINITION 6.1. Let

$$\mathcal{D} = \{\mathfrak{p} \in \mathcal{P} : \Gamma_{S_{\mathfrak{p}}} \text{ is discrete, non-elementary, and torsion-free}\}.$$

and

$$\mathfrak{D} = \{\mathfrak{p} \in \mathcal{D} : (S_{\mathfrak{p}}, U_{\mathfrak{p}}, \mathfrak{Dicta}) \text{ is equivalent to a geometric marking}\}.$$

As alluded to previously, the delicate distinction between $\mathfrak{D}$ and $\mathcal{D}$ is that, in general, even if $\Gamma_{S_\mathfrak{p}}$ is discrete, there could be an element $h \in \Gamma_{S_\mathfrak{p}}$ that violates some condition in $\mathfrak{Dicta}$. In Example CUSPS, recall that a violating $h \in \Gamma_{S_\mathfrak{p}}$ could be one where the interiors of the horoballs $h \cdot U_\mathfrak{p}$ and $U_\mathfrak{p}$ intersect, or h is parabolic and shorter than $m_\mathfrak{p}$. Thus, $\mathcal{D} \setminus \mathfrak{D}$ is comprised of *incorrect* markings. In particular, this implies that for any $\mathfrak{p} \in \mathcal{D} \setminus \mathfrak{D}$ there is $\mathfrak{q} \in \mathfrak{D}$ such that $\Gamma_{S_\mathfrak{p}} = \Gamma_{S_\mathfrak{q}}$, allowing us to eliminate/ignore the parameter $\mathfrak{p}$.

Recall that $\mathfrak{Dicta}$ must also be chosen such that if a condition fails, then we can computationally verify this fact in finite time. As such, the conditions in $\mathfrak{Dicta}$ that cannot be encoded into the parametrization itself must be verifiable by evaluating some inequality. Regions where the inequality fails can then be eliminated from $\mathcal{P}$. Since we are working with groups, our $\mathfrak{Dicta}$ conditions can usually be falsified by finding violating elements $h \in \Gamma_{S_\mathfrak{p}}$, just like in the example above. Of course, one does not need to be restricted to only such criteria. For example, the bound on maximal cusp volume in $\mathcal{P}_{\text{CUSPS}}$ could be thought of as a $\mathfrak{Dicta}$ criterion. However, we will focus on word-related elimination criteria here as they are quite special to this type of parameter space analysis.

6.1. Killer words. Since $S_\mathfrak{p} = \{g_1, \dots, g_k\}$ is an explicit set of generators for $\Gamma_{S_\mathfrak{p}}$ in our Lie group G, it is best to consider words in the free group $\mathfrak{F}$ on symbols $\{t_1, \dots, t_k\}$ and a presentation $\rho_\mathfrak{p} : \mathfrak{F} \to \Gamma_{S_\mathfrak{p}}$ where $\rho_\mathfrak{p}(t_i) = g_i$. We will often abuse notation and use g_i to denote t_i when the context is clear.

DEFINITION 6.2. For a word $w \in \mathfrak{F}$ and $\mathfrak{p} \in \mathcal{P}$, let

$$w_\mathfrak{p} = \rho_\mathfrak{p}(w) = \pm \begin{pmatrix} a_w(\mathfrak{p}) & b_w(\mathfrak{p}) \\ c_w(\mathfrak{p}) & d_w(\mathfrak{p}) \end{pmatrix} \in \mathrm{PSL}(2, \mathbb{C}),$$

where $a_w(\mathfrak{p}), b_w(\mathfrak{p}), c_w(\mathfrak{p}), d_w(\mathfrak{p})$ are the functions assigning matrix entries.

We will use words in $\mathfrak{F}$ to define computable open sets in $\mathcal{P}$ that are guaranteed to be disjoint from $\mathfrak{D}$. Essentially, we look for words that violate conditions specified in $\mathfrak{Dicta}$. If we can give a cover of $\mathcal{P}$ by such open sets, then $\mathcal{P}$ cannot cannot contain any geometric markings. If we are not so lucky, we can at least eliminate large chunks of $\mathcal{P}$ by cutting away these open sets and be left with a few regions of interest. Below are two examples, one of type TUBES and one for CUSPS.

LEMMA 6.3. *Consider markings of type* TUBES *with parameter space* $\mathcal{P}_{\text{TUBES}}$. *Given* $w \in \mathfrak{F}$, *define*

$$\mathcal{S}_w = \{\mathfrak{p} \in \mathcal{P}_{\text{TUBES}} : 0 < \mathrm{re}\operatorname{arccosh}(tr(w_\mathfrak{p})/2) < \mathrm{re}\operatorname{arccosh}(tr(f_\mathfrak{p}))/2)\}.$$

Then $\mathcal{S}_w \cap \mathfrak{D} = \emptyset$.

PROOF. Notice that $\mathcal{S}_w$ is the open set of parameters where the real translation length of $w_\mathfrak{p}$ is nonzero but shorter than that of $f_\mathfrak{p}$. Since the translation length of $w_\mathfrak{p}$ is positive, $w_\mathfrak{p}$ is a loxodromic and so cannot be the identity. If $\mathfrak{p} \in \mathfrak{D}$, this would violates the criterion that $f_\mathfrak{p}$ has shortest real translation length, so we conclude $\mathcal{S}_w \cap \mathfrak{D} = \emptyset$. □

LEMMA 6.4. *In the context of markings of type* CUSPS, *define*

$$\mathcal{Z}_w = \{\mathfrak{p} = (p, s, \ell) \in \mathcal{P}_{\text{CUSPS}} : |c_w(\mathfrak{p})/s| < 1\}$$

and

$$\mathcal{K}_w = \{\mathfrak{p} \in \mathcal{Z}_w : |c_w(\mathfrak{p})| > 0 \text{ or } a_w(\mathfrak{p}) \neq \pm 1 \text{ or } d_w(\mathfrak{p}) \neq \pm 1\}.$$

Then $\mathfrak{D} \cap \mathcal{K}_w = \emptyset$ *for all* $w \in \mathfrak{F}$.

PROOF. Pick $\mathfrak{p} \in \mathcal{K}_w \subset \mathcal{Z}_w$. Let $U_w(\mathfrak{p})$ be the image of the horoball $U_\mathfrak{p}$ under $w_\mathfrak{p}$. Recall that $U_\mathfrak{p}$ has Euclidean height $1/|s|$. Thus, by 1, $U_w(\mathfrak{p})$ has Euclidean height $|s|/|c_w(\mathfrak{p})|^2$. It follows that the interiors of $U_\mathfrak{p}$ and $U_w(\mathfrak{p})$ intersect precisely when $|s|/|c_w(\mathfrak{p})|^2 > 1/|s|$, or equivalently $|c_w(\mathfrak{p})/S| < 1$.

Assume also that $\mathfrak{p} \in \mathfrak{D}$. Then $\Gamma_{S_\mathfrak{p}}$ is geometric and therefore $U_\mathfrak{p}$ is maximal. Since the interiors of $U_w(\mathfrak{p})$ and $U_\mathfrak{p}$ intersect, we must have $U_\mathfrak{p} = U_w(\mathfrak{p})$ by discreteness and therefore $w_\mathfrak{p} \in \langle m_\mathfrak{p}, n_\mathfrak{p} \rangle$. However, since $\mathfrak{p} \in \mathcal{K}_w$, it is impossible for $w_\mathfrak{p}$ to be parabolic fixing ∞, a contradiction. Thus, $\mathfrak{D} \cap \mathcal{K}_w = \emptyset$ for all $w \in \mathfrak{F}$. □

Note, it is possible that $\mathcal{K}_w \cap \mathcal{D}$ is non-empty, but all such points would be incorrectly marked as discussed previously. The upshot of this lemma is that we can eliminate chunks of $\mathcal{P}_{\text{CUSPS}}$ by finding words for which $\mathcal{P}_{\text{CUSPS}} \cap \mathcal{K}_w \neq \emptyset$. We call these words *killer words* for $\mathcal{P}_{\text{CUSPS}}$. For $\mathcal{P}_{\text{TUBES}}$, *killer words* are those w for which $\mathcal{P}_{\text{TUBES}} \cap \mathcal{S}_w \neq \emptyset$. In general, a *killer word* is a word w that can be used to numerically define an open subset of parameters where at least one condition of $\mathfrak{Dicta}$ fails. When choosing the conditions in $\mathfrak{Dicta}$, it is helpful to plan ahead for how one would build such subsets in the given context.

Going back to $\mathcal{P}_{\text{CUSPS}}$, notice that if $\bigcup_{w \in \mathfrak{F}} \mathcal{K}_w$ is a cover of $\mathcal{P}_{\text{CUSPS}}$, then it follows that $\mathfrak{D}$ is empty. In particular, if $\mathcal{P}_{\text{CUSPS}}$ is made a bit smaller by setting the upper bound of $\sqrt{3} - \epsilon$ for the maximal cusp volume, then such a cover exists and is given in **[20]**. However, it is often the case that we are not as lucky. For example, if the cutoff is set at $\sqrt{3} + \epsilon$, then the parameters corresponding to a marking of the figure eight knot complement and its sister will never be covered by $\mathcal{K}_w$ for any w, leaving us with two small regions that we cannot eliminate. One could stop here and have a result that every cusped hyperbolic 3-manifolds has cusp volume greater than $\sqrt{3} + \epsilon$ with the exception of manifolds that contain a marking whose parameter falls into one of these two small regions. Luckily, we can often do better. In the next section, we show how to get further restrictions for the location of the $\mathfrak{D}$ parameters.

6.2. Variety and quasi-relator words. Outside of pure violations of markings, we can make use of many rigidity results about Kleinian groups. The goal here is to extract a list of topological properties that must be satisfied by geometric markings. Our first example is that of the set $\mathcal{Z}_w$ defined for markings of type CUSPS in the previous section.

LEMMA 6.5. *For markings of type* CUSPS, *if* $\mathfrak{p} \in \mathcal{Z}_w \cap \mathfrak{D}$, *then* $m_\mathfrak{p}^a n_\mathfrak{p}^b w_\mathfrak{p} = id$ *for some* $a, b \in \mathbb{Z}$.

PROOF. In the proof of Lemma 6.4, we saw that $w_\mathfrak{p} \in \langle m_\mathfrak{p}, n_\mathfrak{p} \rangle$ because $|c_w(\mathfrak{p})/s| < 1$ forces the interiors of $U_\mathfrak{p}$ and $U_w(\mathfrak{p})$ to intersect and discreteness of $\Gamma_{S_\mathfrak{p}}$ forces them to be equal. It follows that $w_\mathfrak{p} \in \langle m_\mathfrak{p}, n_\mathfrak{p} \rangle$ as we are done. □

If we can cover $\mathcal{P}_{\text{CUSPS}} \setminus \bigcup_{w \in \mathfrak{F}} \mathcal{K}_w$ by a finite collection $\mathcal{Z}_{w_1}, \ldots, \mathcal{Z}_{w_k}$, then we have a lot of restrictions on the groups corresponding to parameters in $\mathfrak{D}$. We call the words $w_1, \ldots, w_k$ *quasi-relator* words. In general, a word is w is a *quasi-relator* if it can be used to numerically define an open subset of parameters where the discrete points admit relators of a given form. In **[20]**, such a cover is found for markings of hyperbolic 3-manifolds with a maximal cusp of volume at most 2.62.

By analyzing the quasi-relator words, the authors prove that all such manifolds were Dehn fillings of a finite list of known manifolds.

Note, in the above example, some extra conditions could be used to force $a = b = 0$. For example, we could require that $w_\mathfrak{p}$ move the point j less than $m_\mathfrak{p}$, which would imply $w_\mathfrak{p} = id$. Thus, one could even find explicit relators using the above technique. Another useful trick that is similar in nature uses the rigidity provided by Jørgensen's and the Shimizu-Leutbecher inequalities.

DEFINITION 6.6. Given $w \in \mathfrak{F}$, define the variety of w by

$$\mathcal{V}_w = \{\mathfrak{p} \in \mathcal{P} : w_\mathfrak{p} = id\}.$$

LEMMA 6.7. *For markings of type* CUSPS, *define the neighborhood $\mathcal{N}_w$ of $\mathcal{V}_w$ by*

$$\mathcal{N}_w = \{\mathfrak{p} \in \mathcal{P} : |c_w(\mathfrak{p})| < 1 \text{ and } |b_w(\mathfrak{p})| < 1\}.$$

For all nontrivial $w \in \mathfrak{F}$, if $\mathfrak{p} \in \mathcal{D} \cap \mathcal{N}_w$ then $\mathfrak{p} \in \mathcal{V}_w$. In particular, w is a relator in $\Gamma_{S_\mathfrak{p}}$.

PROOF. The Shimizu-Leutbecher Theorem [**38**] states that if

$$m_\mathfrak{p} = \pm \begin{pmatrix} 1 & 1 \\ 0 & 1 \end{pmatrix} \quad \text{and} \quad w_\mathfrak{p} = \pm \begin{pmatrix} a & b \\ c & d \end{pmatrix}$$

generate a discrete subgroup of $\mathrm{PSL}(2, \mathbb{C})$ and $|c| < 1$, then we must have $c = 0$. For $\mathfrak{p} \in \mathcal{D}$, $\Gamma_{S_\mathfrak{p}}$ is discrete and torsion-free, in particular, $m_\mathfrak{p}$ and $w_\mathfrak{p}$ generate a discrete subgroup of $\mathrm{PSL}(2, \mathbb{C})$. Suppose $\mathfrak{p} \in \mathcal{D} \cap \mathcal{N}_w$. By definition $|c_w(\mathfrak{p})| < 1$, but discreteness implies $c_w(\mathfrak{p}) = 0$, so $w_\mathfrak{p} \in \langle m_\mathfrak{p}, n_\mathfrak{p} \rangle$. Further, since $|b_w(\mathfrak{p})| < 1$ and $m_\mathfrak{p}$ is a shortest-length generator of $\langle m_\mathfrak{p}, n_\mathfrak{p} \rangle$, we must have $b_w(\mathfrak{p}) = 0$. Therefore, we conclude that $\mathfrak{p} \in \mathcal{V}_w$. □

If we find $w \in \mathfrak{F}$ such that $\mathcal{N}_w \cap \mathcal{P} \neq \emptyset$, then we automatically know that any geometric parameters in that intersection would have w as a relator (and have to lie on $\mathcal{V}_w$). For markings of type CUSPS we call these w a *variety* or *relator* words. In general, a word w is a *variety* or *relator* word if it can be used to numerically define an open subset of parameters where the discrete points must all lie on $\mathcal{V}_w$. While Lemma 6.7 relies on the CUSPS marking, in particular the presence of $m_\mathfrak{p}$ and the fact that it is "shortest," a more general version can be obtained from Jørgensen's inequality. Pick $g \in S$ and let

$$\begin{aligned} \mathcal{J}_w = \{\mathfrak{p} \in \mathcal{P} : & |tr(w_\mathfrak{p})^2 - 4| + |tr(w_\mathfrak{p} g_\mathfrak{p} w_\mathfrak{p}^{-1} g_\mathfrak{p}^{-1}) - 2| < 1 \text{ or} \\ & |tr(g_\mathfrak{p})^2 - 4| + |tr(g_\mathfrak{p} w_\mathfrak{p} g_\mathfrak{p}^{-1} w_\mathfrak{p}^{-1}) - 2| < 1\} \end{aligned}$$

Then for any $\mathfrak{p} \in \mathcal{J}_w$, we must have that $g_\mathfrak{p}$ and $w_\mathfrak{p}$ generate an elementary Kleinian group. In particular, if g is loxodromic, we get a relator of the form $wgw^{-1}g^{-1}$.

Returning back to the setting of CUSPS we can take things a step further. If $\mathfrak{p} \in \mathcal{N}_{w_1} \cap \mathcal{N}_{w_2}$, we know that either $\mathfrak{p}$ arises from geometric markings that have both w_1 and w_2 as relators, or it can be ignored. This technique can now allow us to pinpoint geometric markings exactly as follows. $\mathcal{V}_{w_i}$ corresponds to a variety in $\mathcal{P} \subset \mathbb{C}^d$ whose polynomial equations are encoded in w and use the parameters as variables. The set $\mathcal{V}_{w_1} \cap \mathcal{V}_{w_2} \cap \mathcal{P}$ can often be analyzed by using a computer algebra system, especially if the intersection is a discrete set of points. If we are lucky, the following three thing may happen (1) $\mathcal{V}_{w_1} \cap \mathcal{V}_{w_2} \cap \mathcal{P}$ is just one point, (2) the group with presentation $\langle t_1, \ldots, t_k \mid w_1, w_2 \rangle$ can be identified as the fundamental group of a manifold Y, and (3) Y contains a geometric marking that lands in $\mathcal{P}$. If this is the case, then the intersection point must arise form a geometric marking of Y. Often, this is enough to identify the manifolds on the nose, as is done in [**20**] for the figure eight knot and its sister to show that they have the smallest maximal cusp-volume.

REMARK 6.8. In this section, we mostly gave examples of how *killer*, *quasi-relator*, and *variety/relator* words can arise. The general definitions of these concepts heavily rely on taking a word w and building some open subset of parameters. The method of choosing how to construct such open sets and the algorithms of checking if a parameter lies in one are left to implementation. However, we note that is it important to design $\mathfrak{Dicta}$ in such a way as to make these constructions possible.

7. Computational setup

Now that we have explored the notion of markings, their parameter spaces, and various elimination criteria, our next goal is to describe the computational aspects involved in finding a provable cover of $\mathcal{P}$ by open sets of the forms $\mathcal{K}_w, \mathcal{Z}_w, \mathcal{N}_w$, and similar, depending on the context. The basic idea is to subdivide $\mathcal{P}$ into small boxes and to associate to each box a word w for which we can computationally verify that the entire small box lies in the specified the open set associated to w. Thus, to provide a proof of the existence of such a cover, we need to do three things (1) encode the locations of all the small boxes whose union contains $\mathcal{P}$, (2) find a word and associated open set that covers each box, and (3) to provide verified computational tools to check if a box lands in a given open set. As we shall see, one benefit of this approach will be that the proof can be stored separate from the computational tools and also does not require any complex logic that may have been used to find the words associated to the cover. In fact, as tools for verified computation become better and easier to use, the arithmetic used in [**22**] and [**20**] could easily be replaced while still using the original data. For completeness, we will give a brief discussion of this arithmetic in the last section. Here, we will focus on our methods for accomplishing for (1) and (2).

To encode our subdivision of $\mathcal{P}$ into small boxes, we use two tricks. First, we embed $\mathcal{P}$ into a large bounding box $\mathcal{B}$ that will be easier to subdivide. Then, for the subdivision, we will use the technique of Binary Space Partitioning, which was first used in computer graphics in the 1960's.

Again, we focus on the case of CUSPS, with much of the overview below appearing in [**20**]. We place the compact parameters space $\mathcal{P}_{\text{CUSPS}}$ into a large box

$$\mathcal{B} = \{(x_0, x_1, x_2, x_3, x_4, x_5) \in \mathbb{R}^6 : |x_i| \leq \texttt{scale} \cdot 2^{-i/6}\},$$

where each subsequent side of the box is $1/\sqrt[6]{2}$ times the size of the previous and `scale` is some scaling factor. For $\mathfrak{p} = (p, s, \ell) \in \mathcal{P}_{\text{CUSPS}}$, the embedding is given by $\ell = x_3 + \mathrm{i}x_0$, $s = x_4 + \mathrm{i}x_1$, and $p = x_5 + \mathrm{i}x_2$. In [**20**], `scale` is taken to be 8, which makes $\mathcal{B}$ is large enough to contain all of $\mathcal{P}_{\text{CUSPS}}$ for manifolds of maximal cusp volume less than 2.62.

The constant side ratio of $1/\sqrt[6]{2}$ for $\mathcal{B}$ is chosen specifically such that if we cut along the 1st dimension, then the two resulting boxes have the same constant side ratio as $\mathcal{B}$. Cutting those along the 2nd dimension yields 4 boxes that still have the same constant side ratio. This is reminiscent of the $1/\sqrt{2}$ side ratio of A-series printer paper, except for 6-dimensional boxes instead of rectangles in the plane.

This behavior yields two advantages. First, the (sub-)boxes of $\mathcal{B}$ obtained by cutting in this manner stay relatively "round," making computational corrections for rounding error less dramatic. Second, this subdivision allows us to encode these (sub-)boxes in binary as *boxcodes*. For example, a boxcode 0 corresponds to the box $\mathcal{B}_0$ obtained by cutting $\mathcal{B}$ in half along the 1st dimension and taking the resulting box on the "left." We fix some preferred orientation on $\mathbb{R}^6$ to define "right" and "left." The box $\mathcal{B}_{01}$ corresponds to cutting $\mathcal{B}_0$ in half along the 2nd dimension and taking the piece on the "right." For deeper boxes, we keep cutting along the next dimension and after cutting along the 6th dimension, we start again with the 1st. For a boxcode $\mathfrak{b}$, we let $\mathcal{B}_\mathfrak{b}$ denote the corresponding box. See Figure 1 for a picture of this subdivision for a box in $\mathbb{R}^3$.

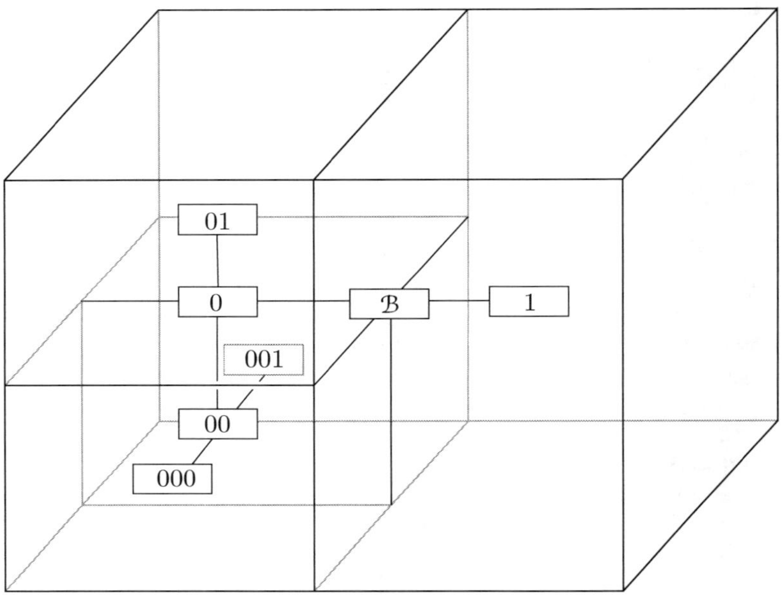

FIGURE 1. Boxcodes and subdivision in $\mathbb{R}^3$ with labels at the box centers from [**22**].

To analyze all of $\mathcal{B}$, we build a binary tree $\mathcal{T}$ corresponding to the boxcodes. The terminal nodes of the tree give a subdivision of $\mathcal{B}$ into boxes of different sizes. The reason for this approach is that the number of boxes needed can be quite large. In [**20**], the number of boxes was 1,394,524,064. If we were to divide $\mathcal{B}$ into boxes all of the same size, our computation and verification time would exponentially increase. In addition, verification times heavily depend on the number of boxes needed. While the tree in [**20**] can be verified in 4-5 hours, once the number of boxes exceeds several billion, the time should be counted in days. Further, the process of finding this proof tree can take months of computation on a multi-core cluster (e.g. 40 nodes and 5 months of wall-time). Much of this time is spend on the word search. However, once a good bank of words is known, reconstructing the tree is quite fast (on the order of days or weeks). In particular, finding the right words is a complicated balancing act and is described in the next section.

Each terminal node $\mathfrak{b}$ of this tree has an associated killer word, quasi-relator word, variety word, boundary condition, or other desired elimination criterion. Here, a *boundary condition* is simply a test that implies that the box does not intersect the embedding of $\mathcal{P}$ in $\mathcal{B}$. In the $\mathcal{P}_{\text{CUSPS}}$ example, this involves checking that one of the inequalities defining $\mathcal{P}_{\text{CUSPS}}$ fails over the entire box. Similarly, if w is a killer word associated to $\mathfrak{b}$, then we can rigorously verify that $\mathcal{B}_{\mathfrak{b}} \subset \mathcal{K}_w$ by evaluating a series of inequalities encoded by w. These conditions could theoretically be checked by hand given an unreasonably large amount of time.

The verification process simply checks conditions at all terminal nodes by traversing the binary tree from the root node (in depth-first order). A successful traversal of the tree guarantees that the terminal boxes form a cover of $\mathcal{B}$ and therefore $\mathcal{P}$. Since each terminal box either misses $\mathcal{P}$ or lies inside an associated open set $\mathcal{K}_w, \mathcal{Z}_w, \mathcal{N}_w$, or similar, we have a good cover of $\mathcal{P}$ that provides information about $\mathfrak{D}$. Example code of a verification program can be found alongside [**22**] and [**20**]. Further, [**22**] goes through specific examples in the text that provide a good understanding of the verification process.

8. Finding useful words

Here, we outline our methods for successfully finding useful words that allow us to obtain a complete decomposition of $\mathcal{B}$ into (sub-)boxes with associated words.

A naive approach uses diagonal enumeration to combine breadth-first enumeration of the tree of boxes with breadth-first enumeration of the Cayley graph on the generators. The naive algorithm has running time $O(3^L 2^D)$, where L is maximum word length and D is maximum box depth. This is much too slow.

To speed it up, there are three basic approaches, all of which are necessary:

(1) avoid considering most boxes by stopping once we have a solution
(2) reuse words that work on one box elsewhere
(3) use geometric heuristics to prefer words that are more likely to work.

The exposition will proceed in rough chronological order, in the hope that by describing some of the wrong turns, we'll help others avoid making the same mistakes. Note, we will mostly focus on killer words because they tend to also work as relator and variety words. In fact, you can see in the examples given, killer words are almost always relator words that just also prove the word is not a relator in the box. However, we often want our relator or variety words to be few and to have special properties, so killer words are still necessary.

The most obvious way of speeding up the search is to avoid the search entirely when feasible: a killer word works on a neighborhood of a region, and by testing killer words found for nearby boxes, most of the time the search is not necessary.

Still, some searches have used words as long as 44, and testing all of the roughly 3^{44} combinations would be prohibitive. Rather than blindly selecting words in first-in-first-out order, the algorithm can rank the words under consideration based on a heuristic estimate of the likelihood of their being useful. We note first that short words tend to be better than long words, as they have fewer steps and less error accumulation across computations. Second, we want words that we expect to violate conditions in $\mathfrak{Dicta}$. This usually involves some geometric measurement about how far a word moves the set U for parameters in a box. For example, the algorithm for the search in [**20**] for markings of type CUSPS was derived from code that would find all w (of bounded word length) for which the horoball $w \cdot U$ has center in a fundamental domain of $\langle m, n \rangle$ and whose Euclidean height is greater than some cutoff δ. Often, this would quickly find a ball that would crash into U. The algorithm can be found as part of visualizing software at [**27**], but is a fun exercise on its own. We describe this algorithm at the end of this section. In summary, the ranking of a word w for a box was a combination of the word length and how close the image of $w \cdot U$ was to U over the box. Such an algorithm, of course, heavily depends on the geometry of the marking. Further, if different types of words are required, mixing several geometric search algorithms could be considered.

Note, this approach can get stuck. If the box is close to a variety, you will often keep finding the same variety word over and over. In some cases, one variety word may be enough, but not in others. For example, the relator you keep finding is not of the form you need or you may need two or more relators to identify a specific group. To step around this, one needs a "diversity" heuristic, which asks for words of a different type, or just a list of "bad relators." For example, if a relator that has been found has a particular abelianization, then look for relators with different abelianizations to guarantee that they are independent.

One other important aspect about searching for and evaluating words: don't use rigorous arithmetic until you think you should. Whenever a word was evaluated, a kind of triage should be used to determine whether that word is likely to kill the box in question, likely to kill any of its nth generation descendants, or unlikely to kill any descendants of the box. An answer to such a heuristic tells us whether to try a verified evaluation (with the error term included), defer further evaluation until the box had been subdivided n more times, or exclude that word from further consideration on any descendant of the box. As simple example of such a heuristic is to use the fastest arithmetic you have to just evaluate at the center of the box. Then, if a words works for the center, check the corners. Finally, if it works on all the corners, try using expensive verified arithmetic over the entire box. However, if the value at the center is too far from being successful for the size of the box (since words encode inequalities so there is a natural measure of "too far"), then the word is probably not going to work for any descendants. And, of course, if the word seems close to working, it might be good to have it at the top of the list for evaluation on smaller sub-boxes. With these heuristics, the program for [**22**] wound up using expensive arithmetic, on average, on about 10 of the roughly 13200 word-bank words per box.

Even with such heuristics in mind, it is important to have control of the word search to prevent it from from running forever. It should be temporarily abandoned after some number of steps, and re-done once the number of descendant boxes doubles or is sufficiently large by some other metric. This way, the search could run forever, but only if the subdivision process runs forever. Knowing how to profile your code to see where the computation spends most of its time is also quite useful. This merged process of alternately searching and subdividing is called the *decomposition algorithm.*

The decomposition algorithm can go through several revisions. One benefit of using the tree data structure for recording successful elimination of boxes is that the search can restart at the boxes that are still open, or it can run over the tree again to make improvements or corrections.

Usually the first attempt — used to determine the feasibility of the whole effort and computational correctness — should do the search in depth-first order on a collection of small sub-boxes of particular interest. These could be boxes around known discrete points or places where simple geometric arguments are hard to do by hand. Such an initial step has two benefits. First, it checks that your arithmetic and word search code works to provide expected results. Second, it seeds a bank of words that can now be used in other parts of the parameter space.

After some amount of feasibility has been determined, a run that uses the existing word bank and the basic boundary conditions can now try to sculpt away the parts of $\mathcal{B}$ that are easy to eliminate. Essentially, leaving parts of $\mathcal{P}$ that aren't close to the killer neighborhoods of the words found in the initial targeted search. Note, this search should run in breadth-first order and do rather little word searching as it is a time consuming process.

In the third stage, one can try a more aggressive word search for the remaining trouble regions, possibly with human input. For example, software that can visualize the action of the group Γ_S on U (or other sets of interest) can be very helpful in finding or even guessing killer words or new conditions that can be added to kill boxes. The search heuristic can also be amended and updated to take into account new observations. Having tools that can provide information about the remaining boxes can also be quite helpful.

If the third stage successfully produces a complete tree, a final step could be do a run that tries to reduce the number of boxes. Essentially, this run attempts all found killer words in a large region (about a thousand boxes) on all the parent boxes in the region and does no new searching.

Finally, once a completed tree is produced, the last task is to produce a minimal program that reads and checks the inequalities encoded in the tree using verified arithmetic. This is important to make sure no bugs in the search code have an effect on tree validation. Further, this allows for including clear and coherent documentation for each test. See [**43**] for an example of such a program.

8.1. Cusps word search algorithm. Recall that when searching for candidate killer or quasi-relator words in the Cusps setup, we are interested in finding all images of the horoball $U = H_\infty$ that have large height. Since we also want to use short words to avoid numerical error growth, we only care about such translates $w \cdot H_\infty$, where w uses few instances of $g^{\pm 1}$. We call this number of instances of $g^{\pm 1}$ the *g-length* of w. For a given parameter $\mathfrak{p}$, our task it to walk in the Cayley graph of $\mathfrak{F}$ and find all w where $w_\mathfrak{p} \cdot H_\infty$ has height at least some number h and w has g-length at most some number K. We start with the horoballs $g^{\pm 1} \cdot H_\infty$. Next, we ask the question: how far can we translate them via $\langle m, n\rangle$ until we know their future images under $g^{\pm 1}$ will have height smaller than h? Indeed, such a bound must exists because the distance between $g \cdot H_\infty$ and $m^a n^b g H_\infty$ will then be the distance between H_∞ and $g^{-1} m^a n^b g H_\infty$. But this latter distance controls the height of $g^{-1} m^a n^b g H_\infty$. A similar distance argument applies to $g^{-1} \cdot H_\infty$ and any translate. Since for each parameter $\mathfrak{p}$ the horoballs $g^{\pm 1} \cdot H_\infty$ are fixed, one can compute a bound on the exponents a, b. Iterating this process allows us to find all horoballs with a given height cutoff which are images of words with given g-length. See Algorithm 1 for an outline given fixed parameter $\mathfrak{p}$.

Algorithm 1: CUSPS word search algorithm

input : Parameter $\mathfrak{p}$, minimum height h, and g-length bound K
output: All words w of g-length $\leq K$ such that $w_{\mathfrak{p}} \cdot H_\infty$ has Euclidean height at least h.

We index a collection of words B by g-length;
depth = 1;
$B[depth] = \{g, g^{-1}\}$;
while *depth* $< K$ **do**
    $B[depth + 1] = \{\}$;
    for w *in* $B[depth]$ **do**
        Find all a, b such for which $(gm^a n^b w)_{\mathfrak{p}} \cdot H_\infty$ or $(g^{-1} m^a n^b w)_{\mathfrak{p}} \cdot H_\infty$ have height at least h;
        Add $gm^a n^b w$ and $g^{-1} m^a n^b w$ to $B[depth + 1]$;
    end
    depth = *depth* + 1;
end
return B

In a discrete group, this process will simply generate disjoint horoballs. However, as soon as one leaves the discrete locus, horoballs will start to intersect. Conjugating one of an intersecting pair to H_∞ guarantees the existence of a large horoball at an indiscrete parameter.

9. Arithmetic and computational considerations

The inequalities in the definitions of $\mathcal{K}_w$, $\mathcal{U}_w$, and $\mathcal{V}_w$ are straightforward to check after constructing the matrix corresponding to w at points of the box $\mathcal{B}_{\mathfrak{b}}$. To prove that these conditions hold over the entire box, we often use affine 1-jets with error and round-off error for computations, which was introduced in [**22**]. If one wanted to use interval arithmetic instead, we expect that further subdivision would be necessary. Using our arithmetic or similar should have significant benefits when implementing a version of the decomposition algorithm. The reason it generally works well is that affine 1-jets approximate a function over the box and not just upper and lower bounds of that function. So if the sum of two functions cancels out and is close to zero, our arithmetic can see this over a large box, while interval arithmetic would need rather fine subdivision.

9.1. 1-jets and error control. Here, we define affine 1-jets in the context of $\mathcal{P}_{\text{CUSPS}}$, which is a 3-complex dimensional parameter space. Let

$$\mathcal{A} = \{(z_0, z_1, z_2) \in \mathbb{C}^3 : |z_i| \leq 1\}$$

and consider a (holomorphic) function $g : \mathcal{A} \to \mathbb{C}$. An affine 1-jet approximation with error ϵ of g is a linear map $(z_0, z_1, z_2) \mapsto c + a_0 z_0 + a_1 z_1 + a_0 z_2$ such that

$$|g(z_0, z_1, z_2) - c + a_0 z_0 + a_1 z_1 + a_0 z_2)| \leq \epsilon.$$

Considering all maps approximated by a given 1-jet, one defines the jet-set

$$S(c, a_0, a_1, a_2; \epsilon) = \{g : \mathcal{A} \to \mathbb{C} : |g(z_0, z_1, z_2) - c + a_0 z_0 + a_1 z_1 + a_0 z_2)| \leq \epsilon\}.$$

In [**22**], the authors derive the arithmetic for jets. For example, given jet-sets S_1 and S_2, they show how to compute the parameters for a jet-set S_3 where $f \cdot g \in S_3$ for all $f \in S_1$ and $g \in S_2$. They do this for all basic arithmetic operations $\pm, \times, /$, and $\sqrt{\cdot}$. If one requires other (holomorphic) operations, new code needs to be written, as well as the accompanying proof of its validity. Note, non-holomorphic arithmetic will require additional parameters as the derivative of a function $f : \mathbb{R}^6 \to \mathbb{R}^2$ is 12 real-dimensional. Code for a version of this arithmetic should be available soon as it is necessary for MARGULIS [**17**].

For our computations in most parameter searches to date, we have used this exact same arithmetic. Note, to make this arithmetic fast, the parameters of $S(c, a_0, a_1, a_2; \epsilon)$ are all given as floating-point doubles. Recall that floating-point numbers are a finite subset of the reals represented via a list of bits on a computer. As such, any arithmetic operation between two floating-point numbers must make a choice about which floating-point result to give, as the real value of this result may not be representable as a floating-point number. On a machine conforming to the IEEE-754 standard [**1**], all operations performed with floating point numbers are guaranteed to round in a consistent way to a closest floating-point representative, as long as overflow and underflow have not occurred. Thus, code for all verification projects must include checks for underflow and overflow during validation, see [**43**] for an example.

In the case of CUSPS, just as in Section 7 of [**22**], each boxcode corresponds to a (sub)box of our parameter space that has a floating-point center $(c_0, c_1, c_2, c_3, c_4, c_5)$ and a floating-point size $(s_0, s_1, s_2, s_3, s_4, s_5)$. Using the IEEE-754 standard, the floating-point size is large enough so that the floating-point version of the box, i.e. $\{(x_0, x_1, x_2, x_3, x_4, x_5) \in \mathbb{R}^6 : |x_i - c_i| \leq s_i\}$, contains the true box as a proper subset. Recall that over a box we associate the coordinate functions $\ell = x_3 + \mathrm{i}x_0$, $s = x_4 + \mathrm{i}x_1$, and $p = x_5 + \mathrm{i}x_2$ for our parameters $\mathfrak{p} = (p, s, \ell) \in \mathcal{P}_{\text{CUSPS}}$. We can replace these coordinate functions with linear maps $g_\ell, g_s, g_p : \mathcal{A} \to \mathbb{C}$ given by:

$$\begin{aligned} g_\ell(z_0, z_1, z_2) &= c_3 + \mathrm{i}c_0 + (s_3 + \mathrm{i}s_0)z_0, \\ g_s(z_0, z_1, z_2) &= c_4 + \mathrm{i}c_1 + (s_4 + \mathrm{i}s_1)z_1, \\ g_p(z_0, z_1, z_2) &= c_5 + \mathrm{i}c_2 + (s_5 + \mathrm{i}s_2)z_2. \end{aligned}$$

Notice that, by construction, the linear maps see all the ℓ, s, p values over the given box. This means that for every point $(p, s, \ell) \in \mathcal{B}_\mathfrak{b}$, there are z_0, z_1, z_2 such that $\ell = g_\ell(z_0, z_1, z_2)$, $s = g_s(z_0, z_1, z_2)$ and $p = g_p(z_0, z_1, z_2)$. In fact, these functions also see values that are a little outside of $\mathcal{B}_\mathfrak{b}$ due to the change from a box in $\mathbb{R}^6$ to a box in $\mathbb{C}^3$. We can now think of g_ℓ, g_s, g_p as living in jet-sets. For example, $g_p \in S(c_5 + \mathrm{i}c_2, 0, 0, s_5 + \mathrm{i}s_2; 0)$ and similarly for others. It follows that any evaluations with jet-arithmetic will include the true values over the box up to the error that accumulates. See [**22**], Sections 7 and 8 for concrete details. Lastly, it is important to remember that this verified arithmetic should be used sparingly: specifically, only when the chances of a successful elimination of a box are high. This is especially important if the dimensionality of the parameter search increases and the cost of this arithmetic grows polynomially in dimension for most operations.

10. Sanity checking

One of the benefits of keeping the verification code separate from the search code is that the latter is prone to frequent changes which may introduce errors. Making sure that the validation is done by a concise, clean, and easily-readable piece of code is essential. For a deeper discussion on why one should be confident in such a `verify` program and its internal consistency, we point the reader to the introduction of [**22**, p 339-341] that covers many aspects of this question in detail. However, during the search process itself, it is also important to preform sanity checks and to build tools that allow by-hand analysis of found examples consistent with theory. To accomplish this, we use several strategies: (1) find boxcodes for as many known examples as one can and validate that the search behaves as expected at those boxcodes, (2) write visualization tools to analyze marking, (3) use other tools or hands-on methods to verify the output of the search, and (4) write and re-run tests of the arithmetic whenever changes to it are made.

In many of our contexts, the first task can be accomplished by using a census of known manifolds. For finite-volume hyperbolic 3-manifolds and knot exteriors, SnapPy [**16**] contains several large collections tabulated by Hodgson, Weeks, Callahan, Dean, Weeks, Champanerkar, Kofman, Patterson, Dunfield, and Christy. Extracting the relevant invariants from the census can be a challenge in its own right, but having concrete boxcodes to test against is essential. See [**44**] for an example. The search must correctly classify all known examples and discrepancies discovered must be thoroughly investigated. Such discrepancies could include things like mistakes in the census-to-boxcode mapping or making incorrect assumptions or generalizations when experimenting with elimination criteria. For example, a census parameter might land on the boundary of a box, so all boxes that contain this point on its boundary must be considered. We must also remark that a given census manifold may give rise to multiple points in the parameter space as the manifolds might contains several valid markings (in TUBES or CUSPS, there might be multiple self-tangencies of a tube or a cusp, giving different markings). In addition, several representatives of the same equivalence class may exist in the parameter space, especially on its boundary.

Another useful approach is the creation of visualization tools. There are two important types to consider. First, given a marking $(S, U, \mathfrak{Dicta})$, it is a good idea to write code that draws the orbit of U under Γ_S alongside any other relevant information. For CUSPS, as studied in [**20**], examples of such code can be found at [**44**], where the cusp diagram of a given marking is drawn from its boxcode. Studying such a visual diagram often helps deal with troublesome boxcodes where the word search algorithm gets stuck, but human intervention can quickly find relator or killer words. Another visualization tool that can be useful is the parameter space itself. Plotting slices through the parameter space and looking for the location of census manifolds and troublesome boxes can give hints to the structure of the set of discrete groups in question.

It is also important to have a by-hand or an alternative methods of building confidence in the partial results as one implements the search. In the context of Kleinian groups, it is often possible to use a combination of software like `heegaard` [**9**] , `twister` [**8**], SnapPy [**16**] and Regina [**12**] to check that relator words found at census boxcodes do indeed recover the marking subgroup. This process goes from a (realizable) presentation to a Heegaard splitting, to a triangulation, and then checks isometry type in SnapPy. This pipeline has been assembled in a small program called `coover` by Haraway [**28**] with code contributions from Dunfield, Linton, Futer, Purcell, Schleimer, and Yarmola. While this tool is not a rigorous method of validation, as compared to the `verify` program, it has played an essential role in finding and correcting issues with more complex elimination criteria and sanity-checking many runs of the search algorithm.

References

[1] *IEEE standard for binary floating-point arithmetic*, ANSI/IEEE Std 754-1985 (1985), 1–20.

[2] Colin C. Adams, *The noncompact hyperbolic 3-manifold of minimal volume*, Proc. Amer. Math. Soc. **100** (1987), no. 4, 601–606, DOI 10.2307/2046691. MR894423

[3] Ian Agol, *Bounds on exceptional Dehn filling*, Geom. Topol. **4** (2000), 431–449, DOI 10.2140/gt.2000.4.431. MR1799796

[4] Ian Agol, *Volume change under drilling*, Geom. Topol. **6** (2002), 905–916, DOI 10.2140/gt.2002.6.905. MR1943385

[5] Ian Agol, *Bounds on exceptional Dehn filling II*, Geom. Topol. **14** (2010), no. 4, 1921–1940, DOI 10.2140/gt.2010.14.1921. MR2680207

[6] Ian Agol, Marc Culler, and Peter B. Shalen, *Dehn surgery, homology and hyperbolic volume*, Algebr. Geom. Topol. **6** (2006), 2297–2312, DOI 10.2140/agt.2006.6.2297. MR2286027

[7] Ian Agol, Peter A. Storm, and William P. Thurston, *Lower bounds on volumes of hyperbolic Haken 3-manifolds*, J. Amer. Math. Soc. **20** (2007), no. 4, 1053–1077, DOI 10.1090/S0894-0347-07-00564-4. With an appendix by Nathan Dunfield. MR2328715

[8] Mark Bell, Tracy Hall, and Saul Schleimer, *Twister (computer software)*, `https://bitbucket.org/Mark_Bell/twister/`, 2008–2014, Version 2.4.1.

[9] John Berge, *heegaard3*, Available at `https://github.com/3-manifolds/heegaard3`, (2021-12-23).

[10] Steven A. Bleiler and Craig D. Hodgson, *Spherical space forms and Dehn filling*, Topology **35** (1996), no. 3, 809–833, DOI 10.1016/0040-9383(95)00040-2. MR1396779

[11] Steven Boyer, *Dehn surgery on knots*, Handbook of geometric topology, North-Holland, Amsterdam, 2002, pp. 165–218. MR1886670

[12] Benjamin A. Burton, Ryan Budney, William Pettersson, et al., *Regina: Software for low-dimensional topology*, `http://regina-normal.github.io/`, 1999–2021.

[13] Chun Cao and G. Robert Meyerhoff, *The orientable cusped hyperbolic 3-manifolds of minimum volume*, Invent. Math. **146** (2001), no. 3, 451–478, DOI 10.1007/s002220100167. MR1869847

[14] David Coulson, Oliver A. Goodman, Craig D. Hodgson, and Walter D. Neumann, *Computing arithmetic invariants of 3-manifolds*, Experiment. Math. **9** (2000), no. 1, 127–152. MR1758805

[15] Thomas Crawford, *A Stronger Gordon Conjecture and an Analysis of Free Bicuspid Manifolds with Small Cusps*, ProQuest LLC, Ann Arbor, MI, 2018. Thesis (Ph.D.)–Boston College. MR3818812

[16] Marc Culler, Nathan M. Dunfield, Matthias Goerner, and Jeffrey R. Weeks, *SnapPy, a computer program for studying the geometry and topology of 3-manifolds* Available at `http://snappy.computop.org` (2021-12-23).

[17] David Futer, David Gabai, and Andrew Yarmola, *On the Margulis constant of hyperbolic 3-manifolds*, In progress.

[18] David Futer, Jessica S. Purcell, and Saul Schleimer, *Effective bilipschitz bounds on drilling and filling*, Geom. Topol. **26** (2022), no. 3, 1077–1188, DOI 10.2140/gt.2022.26.1077. MR4466646

[19] David Gabai, *The Smale conjecture for hyperbolic 3-manifolds:* $\mathrm{Isom}(M^3) \simeq \mathrm{Diff}(M^3)$, J. Differential Geom. **58** (2001), no. 1, 113–149. MR1895350

[20] David Gabai, Robert Haraway, Robert Meyerhoff, Nathaniel Thurston, and Andrew Yarmola, *Hyperbolic 3-manifolds of low cusp volume*, 2021.

[21] David Gabai, Marc Lackenby, and Andrew Yarmola, *Bounds on exceptional Dehn surgery*, In preparation.

[22] David Gabai, G. Robert Meyerhoff, and Nathaniel Thurston, *Homotopy hyperbolic 3-manifolds are hyperbolic*, Ann. of Math. (2) **157** (2003), no. 2, 335–431, DOI 10.4007/annals.2003.157.335. MR1973051

[23] David Gabai, Robert Meyerhoff, and Peter Milley, *Minimum volume cusped hyperbolic three-manifolds*, J. Amer. Math. Soc. **22** (2009), no. 4, 1157–1215, DOI 10.1090/S0894-0347-09-00639-0. MR2525782

[24] David Gabai and Maria Trnkova, *Exceptional hyperbolic 3-manifolds*, Comment. Math. Helv. **90** (2015), no. 3, 703–730, DOI 10.4171/CMH/368. MR3420467

[25] Oliver Goodman, *Snap, a computer program for studying arithmetic invariants of hyperbolic 3-manifolds*, Available at `http://www.ms.unimelb.edu.au/~snap`.

[26] C. McA. Gordon, *Dehn filling: a survey*, Knot theory (Warsaw, 1995), Banach Center Publ., vol. 42, Polish Acad. Sci. Inst. Math., Warsaw, 1998, pp. 129–144. MR1634453

[27] Robert Haraway, *Manifold enumeration software of "hyperbolic 3-manifolds of low cusp volume"*, Available at `https://github.com/bobbycyiii/low-cusp-volume` (2021-12-23).

[28] Robert Haraway et al., *Coover: presentations to triangulations*, Available at `https://github.com/bobbycyiii/coover` (2021-12-23).

[29] Craig D. Hodgson and Steven P. Kerckhoff, *Universal bounds for hyperbolic Dehn surgery*, Ann. of Math. (2) **162** (2005), no. 1, 367–421, DOI 10.4007/annals.2005.162.367. MR2178964

[30] Craig D. Hodgson and Steven P. Kerckhoff, *The shape of hyperbolic Dehn surgery space*, Geom. Topol. **12** (2008), no. 2, 1033–1090, DOI 10.2140/gt.2008.12.1033. MR2403805

[31] Marc Lackenby and Robert Meyerhoff, *The maximal number of exceptional Dehn surgeries*, Invent. Math. **191** (2013), no. 2, 341–382, DOI 10.1007/s00222-012-0395-2. MR3010379

[32] T. H. Marshall and G. J. Martin, *Minimal co-volume hyperbolic lattices, II: Simple torsion in a Kleinian group*, Ann. of Math. (2) **176** (2012), no. 1, 261–301, DOI 10.4007/annals.2012.176.1.4. MR2925384

[33] Robert Meyerhoff, *The cusped hyperbolic* 3*-orbifold of minimum volume*, Bull. Amer. Math. Soc. (N.S.) **13** (1985), no. 2, 154–156, DOI 10.1090/S0273-0979-1985-15401-1. MR799800

[34] Robert Meyerhoff, *Sphere-packing and volume in hyperbolic* 3*-space*, Comment. Math. Helv. **61** (1986), no. 2, 271–278, DOI 10.1007/BF02621915. MR856090

[35] Walter D. Neumann and Don Zagier, *Volumes of hyperbolic three-manifolds*, Topology **24** (1985), no. 3, 307–332, DOI 10.1016/0040-9383(85)90004-7. MR815482

[36] Grisha Perelman, *The entropy formula for the ricci flow and its geometric applications*, 2002.

[37] Grisha Perelman, *Ricci flow with surgery on three-manifolds*, 2003.

[38] Hideo Shimizu, *On discontinuous groups operating on the product of the upper half planes*, Ann. of Math. (2) **77** (1963), 33–71, DOI 10.2307/1970201. MR145106

[39] William P. Thurston, *Geometry and topology of three-manifolds*, Lecture notes available at `http://library.msri.org/books/gt3m/` (2021-12-23), 1978.

[40] William P. Thurston, *Three-dimensional geometry and topology. Vol. 1*, Princeton Mathematical Series, vol. 35, Princeton University Press, Princeton, NJ, 1997. Edited by Silvio Levy. MR1435975

[41] Maria Trnkova and Andrew Yarmola, *On maximal systoles of low-genus hyperbolic surfaces*, In progress.

[42] Jeffrey R. Weeks, *Snap*, Available at `http://www.geometrygames.org/SnapPea/index.html`.

[43] Andrew Yarmola, *Verification software for "Hyperbolic 3-manifolds of low cusp volume"*, Available at `https://github.com/andrew-yarmola/verify-cusp` (2021-12-23).

[44] ———, *Visualization software for "Hyperbolic 3-manifolds of low cusp volume"*, Available at `https://github.com/andrew-yarmola/low-cusp-volume` (2021-12-23).

Department of Mathematics, Fine Hall, Princeton University, Princeton, New Jersey 08544
Email address: gabai@math.princeton.edu

Math Department, Maloney Hall, Boston College, Chestnut Hill, Massachusetts 02467
Email address: robert.meyerhoff@bc.edu

Ipvive Inc 26503 Royal Vista Ct. Santa Clarita, California 91351
Email address: njt@ipvive.com

Department of Mathematics, Fine Hall, Princeton University, Princeton, New Jersey 08544
Email address: yarmola@math.princeton.edu

Contemporary Mathematics
Volume **783**, 2023
https://doi.org/10.1090/conm/783/15698

Exploring Lie theory with GAP

Willem A. de Graaf

ABSTRACT. We illustrate the Lie theoretic capabilities of the computational algebra system GAP4 by reporting on results on nilpotent orbits of simple Lie algebras that have been obtained using computations in that system. Concerning reachable elements in simple Lie algebras we show by computational means that the simple Lie algebras of exceptional type have the Panyushev property. We computationally prove two propositions on the dimension of the abelianization of the centralizer of a nilpotent element in simple Lie algebras of exceptional type. Finally we obtain the closure ordering of the orbits in the null cone of the spinor representation of the group $\mathrm{Spin}_{13}(\mathbb{C})$. All input and output of the relevant GAP sessions is given.

1. Introduction

This paper has two purposes. Firstly, it serves to introduce and advertise the capabilities of the computer algebra system GAP4 [**GAP21**] to perform computations related to various aspects of Lie theory. The main objects related to Lie theory that GAP can deal with directly are Lie algebras and related finite structures such as root systems and Weyl groups. But Lie algebras play an important role in the study of the structure and representations of linear algebraic groups. So also the algorithms implemented in GAP can also be used to perform computations regarding those objects.

The second purpose of the paper is to describe the results of three computational projects that I have been involved in. The first of these is the subject of Section 3 and concerns reachable nilpotent orbits in Lie algebras of exceptional type. Let $\mathfrak{g}$ be a semisimple complex Lie algebra and let $e \in \mathfrak{g}$ be nilpotent. By $\mathfrak{g}_e$ we denote the centralizer of e in $\mathfrak{g}$. The element e is said to be *reachable* if $e \in [\mathfrak{g}_e, \mathfrak{g}_e]$. A nilpotent e lies in a so-called $\mathfrak{sl}_2$-triple, which defines a grading on $\mathfrak{g}$. Panyushev [**Pan04**] proposed a characterization of reachable nilpotent elements in terms of this grading; here we call this the Panyushev property of $\mathfrak{g}$. In [**Pan04**] this property was proved for Lie algebras of type A. Yakimova [**Yak10**] showed that the Lie algebras of type B, C, D also have the Panyushev property. In Section 3 we show by calculations in GAP that the simple Lie algebras of exceptional type also have the Panyushev property.

2020 *Mathematics Subject Classification.* Primary 17B45, 20G05.
Key words and phrases. Lie groups, Lie algebras, nilpotent orbits, computational methods.
The author was partially supported by an Australian Research Council grant, identifier DP190100317.

The second project concerns the quotients $\mathfrak{g}_e/[\mathfrak{g}_e, \mathfrak{g}_e]$ where again e is a nilpotent element in a simple complex Lie algebra $\mathfrak{g}$. These play an important role in [**PT14**]. In Section 4 we show that a statement proved in [**PT14**] for the simple Lie algebras of classical type also holds for the exceptional types, albeit with a few explicitly listed exceptions. The results of Sections 3, 4 have also appeared in the arxiv preprint [**Gra13**], without giving the details of the computations.

In Section 5 we look at the null cone of the spinor representation of the group $\mathrm{Spin}_{13}(\mathbb{C})$. The orbits of this group in the null cone were first listed in [**GV78**]. A' Campo and Popov [**DK15**, Example (f), p. 348] observed, also by computational means, that these orbits coincide with the strata of the null cone (and they corrected the dimensions given in [**GV78**]). Here we show how an algorithm given in [**GVY12**] can be extended to this case to obtain the closure ordering of these orbits. We give a simple implementation in GAP and obtain the closure diagram. Furthermore, we use GAP to study the stabilizers of the elements of the null cone.

In this paper we will not give a full introduction into working with Lie algebras in GAP but refer to the reference manual of GAP which can be found on its website, and to the manuals of the various packages that are listed in the next section. The website of GAP also has various introductory materials of a more general nature. The topics that we discuss in this paper all involve semisimple Lie algebras. For a general introduction to the theory of these algebras we refer to the book by Humphreys, [**Hum78**].

We will give all input and output of the GAP sessions. Most commands return very quickly. If a command takes markedly longer then we display the runtime, by using the GAP function `time`; this command displays the runtime in milliseconds, so that a value of, for example, 23345 means 23.3 seconds.

2. Preliminaries

GAP4 [**GAP21**] is an open source computational algebra system. Its mathematical functionality is contained in a "core system" (which consists of a small kernel written in C and a library of functions written in the GAP language) and a rather large number of packages which can be loaded separately. The GAP library has a number of functions for constructing and working with Lie algebras and their representations. For an overview we refer to the reference manual of GAP. Furthermore there are the following packages that deal with various aspects of Lie theory:

- CoReLG [**DFdG20**], for working with real semisimple Lie algebras.
- FPLSA [**GK19**], for dealing with finitely presented Lie algebras.
- LieAlgDB [**CdGSGT19**], which contains various databases of small dimensional Lie algebras.
- LiePRing, [**VLE18**] containing a database and algorithms for Lie p-rings.
- LieRing [**CdGGT19**], for computing with Lie rings.
- NoCK [**BJS$^+$19**], for the computation of Tolzano's obstruction for compact Clifford-Klein forms.
- QuaGroup [**dGGT19a**], for computations with quantum groups.
- SLA [**dGGT19b**], for computations with various aspects of semisimple Lie algebras.
- Sophus [**SGT18**], for computations in nilpotent Lie algebras.

We also mention the package CHEVIE for dealing with groups of Lie type and related structures such as Weyl groups and Iwahori-Hecke algebras. This package is built on GAP 3, not GAP 4. We refer to its website `https://webusers.imj-prg.fr/~jean.michel/chevie/chevie.html` for more information.

The projects discussed in this paper mainly use the GAP core system and the package SLA. In the next two subsections we briefly look at how simple Lie algebras and their modules are constructed in GAP and how SLA deals with nilpotent orbits in simple Lie algebras.

2.1. Simple Lie algebras in GAP. GAP has a function `SimpleLieAlgebra` for creating the simple split Lie algebras over fields of characteristic 0. (The semisimple Lie algebras can be constructed by the function `DirectSumOfAlgebras`.) They are given by a multiplication table with respect to a Chevalley basis (for the latter concept we refer to [**Hum78**, Theorem 25.2]). For the base field we usually take the rational numbers because often the computations with these algebras are entirely rational, that is, require no solutions to polynomial equations. The i-th basis element of such a Lie algebra is written as `v.i`. In the next example we construct the simple Lie algebra of type E_8, a basis of it and two of its elements.

```
gap> L:= SimpleLieAlgebra( "E", 8, Rationals );
<Lie algebra of dimension 248 over Rationals>
gap> b:= Basis(L);; b[123];
v.123
gap> b[2]-3*b[5]+1/7*b[100];
v.2+(-3)*v.5+(1/7)*v.100
```

Such simple Lie algebras come with a lot of data like a Chevalley basis and a root system. Again we refer to the reference manual for more details.

There also is a function for constructing the irreducible modules of a semisimple Lie algebra. Such a module is given by a highest weight, which is a nonnegative integral linear combination of the fundamental weights. This linear combination is just given by its coefficient vector. (The order of the fundamental weights is given by the Cartan matrix of the root system of the Lie algebra.) The action of an element of the Lie algebra on an element of its module is computed by the infix caret operator `^`. In the next example we construct the irreducible 3875-dimensional module of the Lie algebra of type E_8. We see that the computation in GAP takes about 174 seconds. We also compute the action of an element of the Lie algebra on an element of the module.

```
gap> L:= SimpleLieAlgebra( "E", 8, Rationals );;
gap> V:= HighestWeightModule( L, [1,0,0,0,0,0,0,0] ); time;
<3875-dimensional left-module over <Lie algebra of dimension 248
over Rationals>>
174425
gap> bL:= Basis(L);; bV:= Basis(V);;
gap> bL[1]^bV[263];
-1*y112*v0
```

(For an explanation of the notation of the basis elements of these modules we again refer to the reference manual.)

2.2. Nilpotent orbits in GAP. Here we recall some definitions and facts on nilpotent orbits. For more background information we refer to the the book by Collingwood and McGovern ([**CM93**]). Secondly we show how the package SLA deals with nilpotent orbits.

Let $\mathfrak{g}$ be a semisimple Lie algebra over $\mathbb{C}$ (or over an algebraically closed field of characteristic 0). Let G denote the adjoint group of $\mathfrak{g}$; this is the identity component of the automorphism group of $\mathfrak{g}$. An $e \in \mathfrak{g}$ is said to be nilpotent if the adjoint map $\operatorname{ad} e : \mathfrak{g} \to \mathfrak{g}$ is nilpotent. If $e \in G$ is nilpotent then the entire orbit Ge consists of nilpotent elements, and is therefore called a nilpotent orbit.

By the Jacobson-Morozov theorem a nilpotent $e \in \mathfrak{g}$ lies in an $\mathfrak{sl}_2$-triple (f, h, e) (where $[e, f] = h$, $[h, e] = 2e$, $[h, f] = -2f$). Let $\mathfrak{h} \subset \mathfrak{g}$ be a Cartan subalgebra containing h. Let Φ be the root system of $\mathfrak{g}$ with respect to $\mathfrak{h}$. Then there is a basis of simple roots $\Delta = \{\alpha_1, \dots, \alpha_\ell\}$ of Φ, such that $\alpha_i(h) \in \{0, 1, 2\}$. The Dynkin diagram of Δ, where the node corresponding to α_i is labeled $\alpha_i(h)$, is called a weighted Dynkin diagram. It uniquely determines the orbit Ge.

The nilpotent orbits of the simple Lie algebras have been classified, see [**CM93**]. In the SLA package they can be constructed with the command `NilpotentOrbits`. The output is a list of objects that carry some information such as the weighted Dynkin diagram of the orbit and an $\mathfrak{sl}_2$-triple containing a representative. Here is an example for the Lie algebra of type E_7, where we inspect the weighted Dynkin diagram and the third element of an $\mathfrak{sl}_2$-triple of the 37-th orbit (that is, a representative of the nilpotent orbit).

```
gap> L:= SimpleLieAlgebra("E",7,Rationals);;
gap> no:= NilpotentOrbits(L);;
gap> Length(no);
44
gap> WeightedDynkinDiagram( no[37] );
[ 2, 0, 0, 2, 0, 0, 2 ]
gap> SL2Triple( no[37] )[3];
v.8+v.11+v.13+v.15+v.22+v.23+v.24
```

Now we briefly describe the concept of *induced* nilpotent orbit.

A subalgebra of $\mathfrak{g}$ is said to be *parabolic* if it contains a Borel subalgebra (i.e., a maximal solvable subalgebra). Let $\mathfrak{h}$ be a fixed Cartan subalgebra of $\mathfrak{g}$. Let Φ denote the root system of $\mathfrak{g}$ with respect to $\mathfrak{h}$, and let Δ be a fixed set of simple roots. For a root α we denote the corresponding root space in $\mathfrak{g}$ by $\mathfrak{g}_\alpha$. For a subset $\Pi \subset \Delta$ we define $\mathfrak{p}_\Pi$ to be the subalgebra generated by $\mathfrak{h}$, $\mathfrak{g}_{-\alpha}$ for $\alpha \in \Pi$ and $\mathfrak{g}_\alpha$ for all positive roots α. Then $\mathfrak{p}_\Pi$ is a parabolic subalgebra. Furthermore, every parabolic subalgebra is G-conjugate to a subalgebra of the form $\mathfrak{p}_\Pi$.

Let $\mathfrak{p} = \mathfrak{p}_\Pi$ for a subset $\Pi \subset \Delta$. Let $\Psi \subset \Phi$ be the root subsystem that consists of the roots that are linear combinations of the elements of Π. Then $\mathfrak{p} = \mathfrak{l} \oplus \mathfrak{n}$ where $\mathfrak{l}$ is the subalgebra spanned by $\mathfrak{h}$ and $\mathfrak{g}_\alpha$ for $\alpha \in \Psi$. Secondly, $\mathfrak{n}$ is spanned by $\mathfrak{g}_\alpha$ for positive α that do not lie in Ψ. The decomposition $\mathfrak{p} = \mathfrak{l} \oplus \mathfrak{n}$ is called the Levi decomposition of $\mathfrak{p}$ and the subalgebra $\mathfrak{l}$ is called a (standard) Levi subalgebra of $\mathfrak{g}$. We observe that $\mathfrak{l}$ is a reductive Lie algebra. In the sequel nilpotent orbits in Levi subalgebras appear. The definitions of their properties are the obvious analogues of the definitions concerning semisimple Lie algebras.

Now let $\mathfrak{p} \subset \mathfrak{g}$ be a parabolic subalgebra, with Levi decomposition $\mathfrak{p} = \mathfrak{l} \oplus \mathfrak{n}$. Let $L \subset G$ be the connected subgroup of G with Lie algebra $\mathfrak{l}$. Let Le' be a nilpotent orbit in $\mathfrak{l}$. Lusztig and Spaltenstein ([**LS79**]) have shown that there is a unique nilpotent orbit $Ge \subset \mathfrak{g}$ such that $Ge \cap (Le' \oplus \mathfrak{n})$ is open and nonempty in $Le' \oplus \mathfrak{n}$. The orbit Ge is said to be *induced* from the orbit Le'. Nilpotent orbits which are not induced are called *rigid*.

Let n be a non-negative integer. The irreducible components of the locally closed set

$$A^n = \{x \in \mathfrak{g} \mid \dim Gx = n\}$$

are called *sheets* of $\mathfrak{g}$ (see [**Bor82**], [**BK79**]). A sheet is G-stable and contains a *unique* nilpotent orbit. Sheets in general are not disjoint, and different sheets may contain the same nilpotent orbit. The sheets of $\mathfrak{g}$ are indexed by G-classes of pairs $(\mathfrak{l}, Le')$, where $\mathfrak{l}$ is a Levi subalgebra, and Le' is a rigid nilpotent orbit in $\mathfrak{l}$, see [**Bor82**]. The nilpotent orbit that is contained in the corresponding sheet is equal to the nilpotent orbit induced from Le'. The *rank* of the sheet corresponding to the pair $(\mathfrak{l}, Le')$ is defined to be the dimension of the centre of $\mathfrak{l}$.

In the SLA package a sheet is represented by a *sheet diagram*. We first explain how this is defined. Consider a parabolic subalgebra $\mathfrak{p} = \mathfrak{p}_\Pi$ with corresponding Levi subalgebra $\mathfrak{l}$. Let Le' be a rigid nilpotent orbit in $\mathfrak{l}$, then the pair $(\mathfrak{l}, Le')$ corresponds to a sheet. Now we label the Dynkin diagram of Φ in the following way. Write $\Delta = \{\alpha_1, \dots, \alpha_\ell\}$. If $\alpha_i \notin \Pi$ then node i has label 2. The subdiagram consisting of the nodes i such that $\alpha_i \in \Pi$ is the Dynkin diagram of the semisimple part of $\mathfrak{l}$. To these nodes we attach the labels of the weighted Dynkin diagram of Le'. It is known that the weighted Dynkin diagram of a rigid nilpotent orbit only has labels 0,1. So from a sheet diagram we can identify $\mathfrak{l}$ and Le' and hence the corresponding sheet.

The SLA package has a function `InducedNilpotentOrbits` for computing the induced nilpotent orbits of a simple Lie algebra. This function returns a list of records that is in bijection with the sheets of the Lie algebra. Each record has two components: `norbit` which is the nilpotent orbit contained in the sheet, and `sheetdiag` which is the list of labels of the sheet diagram of the sheet. Here is an example for the simple Lie algebra of type E_7.

```
gap> L:= SimpleLieAlgebra( "E", 7, Rationals );;
gap> ind:= InducedNilpotentOrbits( L );;
gap> Length( ind );
46
gap> ind[34];
rec( norbit := <nilpotent orbit in Lie algebra of type E7>,
  sheetdiag := [ 2, 0, 0, 1, 0, 2, 2 ] )
gap> WeightedDynkinDiagram( ind[19].norbit );
[ 0, 0, 0, 2, 0, 0, 2 ]
gap> WeightedDynkinDiagram( ind[22].norbit );
[ 0, 0, 0, 2, 0, 0, 2 ]
gap> WeightedDynkinDiagram( ind[34].norbit );
[ 0, 0, 0, 2, 0, 0, 2 ]
```

The numbering of the nodes of the Dynkin diagram of the Lie algebra of type E_7 follows [**Hum78**, §11.4]. Hence the sheet diagram of the 34-th sheet is

$$\begin{matrix} & & 0 & & & \\ 2 & 0 & 1 & 0 & 2 & 2 \end{matrix}$$

We obtain the Dynkin diagram of the corresponding Levi subalgebra $\mathfrak{l}$ by removing the nodes labeled 2; wee see that its semisimple part is of type D_4. The weighted Dynkin diagram of the rigid nilpotent orbit in $\mathfrak{l}$ has a 1 on the central node and zeros elsewhere. The rank of the sheet is the dimension of the centre of $\mathfrak{l}$; this is the number of 2's in the diagram, and we see that it is 3.

Furthermore we see that sheets 19 and 22 contain the same nilpotent orbit. By inspection it can be verified that there are no other sheets that contain this nilpotent orbit. Hence this is a nilpotent orbit lying in three sheets.

3. Reachable elements

For $e \in \mathfrak{g}$ we denote its centralizer in $\mathfrak{g}$ by $\mathfrak{g}_e$. In [**Pan04**] an e in $\mathfrak{g}$ is defined to be *reachable* if $e \in [\mathfrak{g}_e, \mathfrak{g}_e]$. Such an element has to be nilpotent. It is obvious that e is reachable if and only if all elements in its orbit are reachable. Hence if e is reachable then we also say that its orbit Ge is reachable.

In [**EG93**], Elashvili and Grélaud listed the reachable orbits in simple complex Lie algebras $\mathfrak{g}$ (in that paper reachable elements are called *compact*, in analogy with [**BB92**]). For a given semisimple Lie algebra we can easily obtain this classification in GAP4, using the SLA package. Here is an example for the simple Lie algebra of type E_6.

```
gap> L:= SimpleLieAlgebra( "E", 6, Rationals );;
gap> nL:= NilpotentOrbits( L );;
gap> reach:= [ ];;
gap> for o in nL do
> e:= SL2Triple( o )[3]; ge:= LieCentralizer( L, Subalgebra(L,[e]) );
> if e in LieDerivedSubalgebra( ge ) then Add( reach, o ); fi;
> od;
gap> Length( reach );
6
gap> WeightedDynkinDiagram( reach[3] );
[ 0, 0, 0, 1, 0, 0 ]
```

This simple procedure obtains six reachable nilpotent orbits. For each such orbit we can look at its weighted Dynkin diagram to identify it in the known lists of nilpotent orbits as in [**CM93**, §8.4]. The third element of our list of reachable orbits corresponds to the orbit with label $3A_1$ in the list in [**CM93**, §8.4].

Let $e \in \mathfrak{g}$ be nilpotent, lying in the $\mathfrak{sl}_2$-triple (f, h, e). The subalgebra spanned by (f, h, e) acts on $\mathfrak{g}$ (by restricting the adjoint representation of $\mathfrak{g}$). By the representation theory of $\mathfrak{sl}_2(\mathbb{C})$ the eigenvalues of $\operatorname{ad} h$ are integers. Hence we get a grading

$$\mathfrak{g} = \bigoplus_{k \in \mathbb{Z}} \mathfrak{g}(k)$$

where $\mathfrak{g}(k) = \{x \in \mathfrak{g} \mid [h, x] = kx\}$. Now set $\mathfrak{g}(k)_e = \mathfrak{g}(k) \cap \mathfrak{g}_e$, and let $\mathfrak{g}(\geq 1)_e$ denote the subalgebra spanned by all $\mathfrak{g}(k)_e$, $k \geq 1$.

Panyushev ([**Pan04**]) showed that, for $\mathfrak{g}$ of type A_n, e is reachable if and only if $\mathfrak{g}(\geq 1)_e$ is generated as Lie algebra by $\mathfrak{g}(1)_e$. Here we call this the *Panyushev property* of $\mathfrak{g}$. In [**Pan04**] it is stated that this property also holds for the other classical types and the question is posed whether it holds for the exceptional types. In [**Yak10**] a proof is given that the Panyushev property holds in types B_n, C_n, D_n. Computations in GAP show that it also holds for the exceptional types.

PROPOSITION 3.1. *Let $\mathfrak{g}$ be a simple Lie algebra of exceptional type. Then $\mathfrak{g}$ has the Panyushev property.*

PROOF. One direction is easily seen to hold in general. Indeed, suppose that if $\mathfrak{g}(\geq 1)_e$ is generated as Lie algebra by $\mathfrak{g}(1)_e$. Since $e \in \mathfrak{g}(2)_e$ it immediately follows that e is reachable.

The converse is shown by case by case computations in GAP. Here we show this for the Lie algebra of type E_6. We let `reach` be the list of reachable nilpotent orbits, as computed above.

```
gap> for o in reach do
> e:= SL2Triple( o )[3]; ge:= LieCentralizer( L, Subalgebra(L,[e]) );
> h:= SL2Triple( o )[2]; gr:= SL2Grading( L, h );
> gegeq1:= Intersection( ge, Subspace( L, Concatenation( gr[1] ) ) );
> ge1:= Intersection( ge, Subspace( L, gr[1][1] ) );
> Print( Subalgebra( L, Basis(ge1) ) = gegeq1, " " );
> od;
true true true true true true
```

The identifier `gr` contains the grading corresponding to the $\mathfrak{sl}_2$-triple. This is a list consisting of three lists. The first of these has bases of the subspaces $\mathfrak{g}(1)$, $\mathfrak{g}(2)$, ... So $\mathfrak{g}(\geq 1)_e$ is the intersection of $\mathfrak{g}_e$ and the subspace spanned by all elements in the union of the lists in `gr[1]`; this space is assigned to the identifier `gegeq1`. Secondly, $\mathfrak{g}(1)_e$ is the intersection of $\mathfrak{g}_e$ and the subspace spanned by the first element of `gr[1]`; this space is assigned to `ge1`. The penultimate line instructs GAP to print `true` if the subalgebra generated by $\mathfrak{g}(1)_e$ is equal to $\mathfrak{g}(\geq 1)_e$. □

Yakimova ([**Yak10**]) studied the stronger condition $\mathfrak{g}_e = [\mathfrak{z}_e, \mathfrak{g}_e]$. In this paper we call elements e satisfying this condition *strongly reachable*. She showed that for $\mathfrak{g}$ of classical type, e is strongly reachable if and only if the nilpotent orbit of e is rigid. By an explicit example this is shown to fail for $\mathfrak{g}$ of exceptional type. For the exceptional types we can show the following.

PROPOSITION 3.2. *Let $\mathfrak{g}$ be a simple Lie algebra of exceptional type. Let $e \in \mathfrak{g}$ be nilpotent. Then e is strongly reachable if and only if e is both reachable and rigid.*

PROOF. If e is strongly reachable then it is reachable, but also rigid by [**Yak10**], Proposition 11. As the SLA package has a function for determining the rigid nilpotent orbits, the converse can easily be shown by direct computation. But it also follows from Proposition 3.1. Indeed, if e is rigid then $\mathfrak{g}(0)_e$ is semisimple, so $[\mathfrak{g}(0)_e, \mathfrak{g}(0)_e] = \mathfrak{g}(0)_e$. Furthermore, $[\mathfrak{g}(0)_e, \mathfrak{g}(1)_e] = \mathfrak{g}(1)_e$ by [**Yak10**, Lemma 8] (where this is shown to hold for all nilpotent e). By the Panyushev property this implies that $[\mathfrak{g}_e, \mathfrak{g}_e] = \mathfrak{g}_e$. □

REMARK 3.3. We can easily compute the rigid nilpotent orbits that are not strongly reachable. Here is an example for the Lie algebra of type E_8.

```
gap> L:= SimpleLieAlgebra( "E", 8, Rationals );;
gap> rig:= RigidNilpotentOrbits( L );;
gap> exc:= [ ];;
gap> for o in rig do
> e:= SL2Triple( o )[3]; ge:= LieCentralizer( L, Subalgebra(L,[e]) );
> if ge <> LieDerivedSubalgebra(ge) then Add( exc, o ); fi;
```

```
> od;
gap> Length( exc );
3
gap> WeightedDynkinDiagram( exc[1] );
[ 0, 0, 0, 0, 0, 1, 0, 1 ]
```

We see that we have obtained three nilpotent orbits that are rigid but not strongly reachable. Comparing the weighted Dynkin diagram of the first of those orbits with the tables in [**CM93**] we see that its Bala-Carter label is A_3+A_1. Table 1 contains the rigid but not strongly reachable orbits in the Lie algebras of exceptional type; it is used in the proof of [**PS18**, Lemma 3.7]. For an explanation of the notation used for the labels we refer to [**CM93**, §8.4].

TABLE 1. Rigid but not strongly reachable nilpotent orbits

type	E_7	E_8	E_8	E_8	F_4	G_2
label	$(A_3+A_1)'$	A_3+A_1	$D_5(a_1)+A_2$	A_5+A_1	$\widetilde{A}_2+A_1$	A_1
$(\dim \mathfrak{g}_e, \dim[\mathfrak{g}_e,\mathfrak{g}_e])$	(41,40)	(84,83)	(46,45)	(46,45)	(16,15)	(6,5)

From the last line we see that in all cases $[\mathfrak{g}_e,\mathfrak{g}_e]$ is of codimension 1 in $\mathfrak{g}_e$. Taking Proposition 3.2 into account we see that this implies that $\mathfrak{g}_e = \langle e\rangle \oplus [\mathfrak{g}_e,\mathfrak{g}_e]$. In [**PS18**] the e with this property are called *almost reachable.*

4. The quotients $\mathfrak{c}_e$

Let $\mathfrak{g}$ be a simple Lie algebra, and e a representative of a nilpotent orbit. As before we denote its centralizer by $\mathfrak{g}_e$. In this section we consider the quotient $\mathfrak{c}_e = \mathfrak{g}_e/[\mathfrak{g}_e,\mathfrak{g}_e]$. These have been studied by Premet and Topley [**PT14**] in relation to finite W-algebras. In [**PT14**] it is shown that the statement of Proposition 4.1 holds without exceptions for the classical Lie algebras. Proposition 4.1, as well as the tables of [**Gra13**, Section 3], are used in [**PT14**] for showing that for $\mathfrak{g}$ of exceptional type, $U(\mathfrak{g},e)^{\mathrm{ab}}$ (the abelianization of a finite W-algebra $U(\mathfrak{g},e)$) is isomorphic to a polynomial ring (with the same six exceptions as Proposition 4.1).

PROPOSITION 4.1. *Let $\mathfrak{g}$ be a simple Lie algebra of exceptional type. Let $e\in\mathfrak{g}$ be a representative of an induced nilpotent orbit lying in a unique sheet. Then the rank of that sheet is equal to* $\dim \mathfrak{c}_e$, *except the cases listed in Table* 2.

TABLE 2. Table of exceptions to Proposition 4.1.

$\mathfrak{g}$	label	weighted Dynkin diagram	rank	$\dim \mathfrak{c}_e$
E_6	A_3+A_1	1 0 1 0 1 0	1	2
E_7	$D_6(a_2)$	1 0 1 0 1 0 2	2	3
E_8	$D_6(a_2)$	1 0 1 0 0 0 1 0	1	3
E_8	$E_6(a_3)+A_1$	0 1 0 0 1 0 1 0	1	3
E_8	$E_7(a_2)$	1 0 1 0 1 0 2 2	3	4
F_4	$C_3(a_1)$	1 0 1 0	1	3

PROOF. The proof is obtained by explicit computations in GAP with the SLA package loaded. We show the computation for the Lie algebra of type E_8. First we compute the list of sheets (as explained in Section 2.2). For each sheet we compute $\dim \mathfrak{c}_e$, where e is a representative of the unique nilpotent orbit in the sheet. These dimensions are stored in the list dims. Secondly, for each sheet we compute the number of sheets having the same nilpotent orbit as the given sheet. This number is stored in the list nr.

```
gap> L:= SimpleLieAlgebra( "E", 8, Rationals );;
gap> shts:= InducedNilpotentOrbits( L );;
gap> nr:= [ ];; dims:= [ ];;
gap> for s in shts do
> e:= SL2Triple( s.norbit )[3];
> ge:= LieCentralizer( L, Subalgebra( L, [e] ) );
> Add( dims, Dimension(ge)-Dimension(LieDerivedSubalgebra(ge)) );
> Add( nr, Length( Filtered( shts, t -> t.norbit = s.norbit ) ) );
> od;
```

For each sheet whose nilpotent orbit lies in no other sheet (that is, the corresponding element of nr is 1) we compute its rank, which is equal to the number of 2's in its sheet diagram (see Section 2.2). If the rank is not equal to $\dim \mathfrak{c}_e$ then we store the sheet in the list exc. At the end this list contains the elements of Table 2.

```
gap> exc:= [ ];;
gap> for i in [1..Length(shts)] do
> if nr[i]=1 then
> rk:= Length( Filtered( shts[i].sheetdiag, x -> x = 2 ) );
> if rk <> dims[i] then Add( exc, shts[i] ); fi;
> fi; od;
gap> WeightedDynkinDiagram( exc[1].norbit );
[ 0, 1, 1, 0, 1, 0, 2, 2 ]
gap> Length( Filtered( exc[1].sheetdiag, x -> x = 2 ) );
3
gap> Position( shts, exc[1] );
8
gap> dims[8];
4
```

So we have obtained the data of the penultimate line of Table 2. □

PROPOSITION 4.2. *Let $\mathfrak{g}$ be a simple Lie algebra of exceptional type, and let $e \in \mathfrak{g}$ be a nilpotent orbit that lies in more than one sheet. Then the maximal rank of such a sheet is strictly smaller than* $\dim \mathfrak{c}_e$.

PROOF. Also this proposition is proved by direct computation. Again we show the computation for the simple Lie algebra of type E_8. We assume that the first part of the computation explained in the proof of the previous proposition has been done. In this case, for each sheet such that the corresponding number in nr is greater than one, we first determine all sheets that have the same nilpotent orbit. This list is assigned to the identifier sh. Then we compute the rank of all sheets in sh. If the maximum of those ranks is not strictly smaller than $\dim \mathfrak{c}_e$ (where e is a representative of the nilpotent orbit in the considered sheet) then we print a ?; otherwise we print a !. Since we only obtain !, the proposition is proved in this case.

```
gap> for i in [1..Length(shts)] do
> if nr[i] > 1 then
> sh:= Filtered( shts, t -> t.norbit = shts[i].norbit );
> rks:= List( sh, r -> Length( Filtered( r.sheetdiag, x -> x=2 ) ) );
> if Maximum( rks ) >= dims[i] then Print("?"); else Print("!"); fi;
> fi; od;
!!!!!!!!!!!!!!!!!!!!!!!!!!!!!!!!!!!!!!
```

□

REMARK 4.3. Let $e \in \mathfrak{g}$ be nilpotent lying in the $\mathfrak{sl}_2$-triple (f, h, e). The Jacobi identity implies that the adjoint map $\mathrm{ad}\, h : \mathfrak{g} \to \mathfrak{g}$ stabilizes $\mathfrak{g}_e$ and $[\mathfrak{g}_e, \mathfrak{g}_e]$. Hence it induces a map $\mathrm{ad}\, h : \mathfrak{c}_e \to \mathfrak{c}_e$. The representation theory of $\mathfrak{sl}_2(\mathbb{C})$ implies that $\mathrm{ad}\, h$ acts with non-negative integral eigenvalues on $\mathfrak{c}_e$. The paper [**Gra13**] contains tables listing those eigenvalues for the nilpotent orbits of exceptional simple Lie algebras.

5. Closures of nilpotent orbits of Spin_{13}

Let G be a reductive complex algebraic group and let V be a finite-dimensional rational G-module. Then the invariant ring $\mathbb{C}[V]^G$ is finitely generated by homogeneous elements. The null cone $N_G(V)$ is defined to be the zero locus of the homogeneous invariants of positive degree. The null cone is stable under the action of G but in general consists of an infinite number of orbits. Hesselink [**Hes79**] constructed a stratification of the null cone, by which it is possible to study its geometric properties. In [**VP89**, §5.5, 5.6] Popov and Vinberg gave a version of this theory in characteristic 0 that works with certain elements, called characteristics, in the Lie algebra of G. Popov [**Pop03**] developed an algorithm to compute these characteristics.

The G-module V is said to be *visible* (or *observable*) if the null cone has a finite number of orbits. Kac [**Kac80**] classified the visible representations of reductive algebraic groups. It turns out that irreducible visible representations of connected simple groups either arise as so-called θ-groups or as the spinor modules of $\mathrm{Spin}_{11}(\mathbb{C})$ and $\mathrm{Spin}_{13}(\mathbb{C})$. For an algorithm for determining the closures of the nilpotent orbits of a θ-group we refer to [**GVY12**]. The orbits of the spinor module $\mathrm{Spin}_{11}(\mathbb{C})$ have been determined by Igusa [**Igu70**]. It is likely that the closures of the nilpotent orbits can be determined in the same way as is done below.

Kac and Vinberg [**GV78**] classified the orbits of the group $\mathrm{Spin}_{13}(\mathbb{C})$ on its 64-dimensional spinor module. It turns out that the null cone has 13 orbits (excluding 0). A' Campo and Popov [**DK15**, Example (f), p. 348], using their implementation of Popov's algorithm [**DK15**, Appendix C] for computing the characteristics of the strata, observed that there are also 13 strata in the null cone. This implies that the strata are orbits. Moreover, their computations gave the dimensions of the orbits in the null cone, which were not all correctly given in [**GV78**].

The package SLA also has an implementation of Popov's algorithm. So we can recover these observations by a computation using that package. In this section we give an algorithm, which is similar to an algorithm given in [**GVY12**], to determine when the (Zariski-) closure of a stratum contains a given other stratum. This algorithm works under some hypotheses that are shown to be satisfied by the spinor module of $\mathrm{Spin}_{13}(\mathbb{C})$. We discuss a simple implementation of this algorithm in GAP and we obtain the Hasse diagram of the closures of the orbits in the null cone of the spinor module of $\mathrm{Spin}_{13}(\mathbb{C})$.

5.1. Preliminaries on the strata of the nullcone. Everything we will say here works for reductive groups, but for simplicity we consider a simple algebraic group G over $\mathbb{C}$. We let $\mathfrak{g}$ be its Lie algebra and $(\ ,\ ) : \mathfrak{g} \times \mathfrak{g} \to \mathbb{C}$ the Killing form (so $(x, y) = \mathrm{Tr}((\mathrm{ad}\, x)(\mathrm{ad}\, y))$). We say that a semisimple element $h \in \mathfrak{g}$ is *rational* if the eigenvalues of $\mathrm{ad}\, h$ lie in $\mathbb{Q}$. This is equivalent to saying that the eigenvalues of h on any $\mathfrak{g}$-module are rational.

Let $\mathfrak{h} \subset \mathfrak{g}$ be a Cartan subalgebra. Then by $\mathfrak{h}_{\mathbb{Q}}$ we denote the set of its rational elements, which is a vector space over $\mathbb{Q}$ of dimension $\dim_{\mathbb{C}} \mathfrak{h}$. We define the norm of $h \in \mathfrak{h}_{\mathbb{Q}}$ by $\|h\| = \sqrt{(h,h)}$.

Now we let V be a rational G-module and consider the null cone $N_G(V)$. By the Hilbert-Mumford criterion a $v \in V$ lies in $N_G(V)$ if and only if there is a cocharacter $\chi : \mathbb{C}^* \to G$ such that $\lim_{t \to 0} \chi(t) \cdot v = 0$ (see [**Kra84**, Section III.2]). Setting $h = d\chi(1)$ we have that h is a rational semisimple element and writing v as a sum of h-eigenvectors we get that the corresponding eigenvalues are all positive.

For a rational semisimple $h \in \mathfrak{g}$ and $\tau \in \mathbb{Q}$ we let V_τ be the τ-eigenspace of h. Furthermore, we set

$$V_{\geq 2}(h) = \bigoplus_{\tau \geq 2} V_\tau(h).$$

Let $v \in V$. Then a *characteristic* of v is a shortest rational semisimple element $h \in \mathfrak{g}$ such that $v \in V_{\geq 2}(h)$. We have the following facts concerning characteristics (see [**VP89**, §5.5, 5.6], [**Gra17**, §7.4.1, 7.4.2]):

(1) v has a characteristic if and only if $v \in N_G(V)$.
(2) If $h \in \mathfrak{g}$ is a characteristic of $v \in V$ and $g \in G$ then $\mathrm{Ad}(g)(h)$ is a characteristic of gv.
(3) Let $\mathfrak{h}$ be a fixed Cartan subalgebra of $\mathfrak{g}$. Then there are a finite number of characteristics $h_1, \ldots, h_s$ in $\mathfrak{h}$, up to the action of G.
(4) For $1 \leq i \leq s$ let $S(h_i)$ be the set of all $v \in N_G(V)$ such that v has a characteristic that is G-conjugate to h_i. The set $S(h_i)$ is called the stratum corresponding to h_i.
(5) The stratification of $N_G(V)$ is $N_G(V) = S(h_1) \cup \cdots \cup S(h_s)$ (disjoint union).

Popov [**Pop03**] (see also [**Gra17**, §7.4.3]) devised an algorithm to compute the characteristics $h_1, \ldots, h_s$ in $\mathfrak{h}$. The algorithm also computes the dimensions of the corresponding strata.

5.2. Closures of the strata. The topological notions (closed sets, open sets, closure, ...) that we use here are relative to the Zariski topology.

Let $\mathfrak{h}$ be a fixed Cartan subalgebra of $\mathfrak{g}$. For a rational $h \in \mathfrak{h}$ we let $Z(h) = \{g \in G \mid \mathrm{Ad}(g)(h) = h\}$; then $\mathfrak{z}(h) = \{x \in \mathfrak{g} \mid [x, h] = 0\}$ is the Lie algebra of $Z(h)$. Both $Z(h)$ and $\mathfrak{z}(h)$ stabilize the spaces $V_\tau(h)$ for $\tau \in \mathbb{Q}$.

Let $h_1, \dots, h_s \in \mathfrak{h}$ be the characteristics of the strata of the nullcone of V. Here we assume two things:

(1) Each $V_2(h_i)$ has an open $Z(h_i)$-orbit.
(2) The strata coincide with the G-orbits in the nullcone.

REMARK 5.1. Let h be one of the characteristics. A $v \in V_2(h)$ lies in the open $Z(h)$-orbit if and only if $\mathfrak{z}(h) \cdot v = V_2(h)$.

Under these hypotheses we can generalize a few results from [**GVY12**].

LEMMA 5.2. *Let h be one of the characteristics. Then the open $Z(h)$-orbit in $V_2(h)$ is equal to $V_2(h) \cap S(h)$. Moreover, h is a characteristic of every element in $V_2(h) \cap S(h)$.*

PROOF. Let u be an element of the open $Z(h)$-orbit in $V_2(h)$. From Theorem 5.4 in [**VP89**] it follows that the set of elements of $V_2(h)$ with characteristic h is open and nonempty. As nonempty open sets intersect, there is a $g \in Z(h)$ such that $g \cdot u$ has characteristic h. But then the characteristic of $u = g^{-1} \cdot (gu)$ is $\mathrm{Ad}(g^{-1})(h) = h$. It follows that h is a characteristic of u, and in particular that $u \in S(h)$.

For $\tau \in \mathbb{Q}$ and $w \in V_2(h)$ set

$$\mathfrak{g}_w = \{x \in \mathfrak{g} \mid x \cdot w = 0\}, \text{ and } \mathfrak{g}_{\tau,w} = \{x \in \mathfrak{g}_w \mid [h,x] = \tau x\}.$$

Let $v \in V_2(h) \cap S(h)$. Since v lies in the closure of $Z(h)u$, we have that $\dim \mathfrak{g}_{\tau,v} \geq \dim \mathfrak{g}_{\tau,u}$, for all $\tau \in \mathbb{Q}$. Because $u, v \in S(h)$ and our assumption that the strata are G-orbits, v and u lie in the same G-orbit. Hence $\dim \mathfrak{g}_v = \dim \mathfrak{g}_u$. But $\mathfrak{g}_v$ is the direct sum of the various $\mathfrak{g}_{\tau,v}$, and similarly for $\mathfrak{g}_u$. It follows that $\dim \mathfrak{g}_{\tau,v} = \dim \mathfrak{g}_{\tau,u}$ for all τ. But

$$\mathfrak{g}_{0,v} = \{x \in \mathfrak{z}(h) \mid x \cdot v = 0\},$$

and similarly for $\mathfrak{g}_{0,u}$. This implies that $\dim \mathfrak{z}(h)v = \dim \mathfrak{z}(h)u$. So also the orbit $Z(h)v$ is open in $V_2(h)$ by Remark 5.1. In particular, v lies in the open $Z(h)$-orbit in $V_2(h)$. □

LEMMA 5.3. *Let W denote the Weyl group of the root system of $\mathfrak{g}$. Let h, h' be two of the characteristics. Then $S(h')$ is contained in the closure of $S(h)$ if and only if there is a $w \in W$ such that $U = V_2(h') \cap V_{\geq 2}(wh)$ contains a point of $S(h')$. Furthermore, the intersection of U and $S(h')$ is open in U.*

PROOF. Here we use the fact that $\overline{S(h)} = GV_{\geq 2}(h)$ ([**VP89**], Theorem 5.6). This immediately implies the "only if" part.

Let $P(h)$ denote the parabolic subgroup with Lie algebra $\oplus_{\tau \geq 0} \mathfrak{g}_\tau(h)$. Using the Bruhat decomposition we then have

$$\overline{S(h)} = \bigcup_{w \in W} P(h')wP(h)(V_{\geq 2}(h)) = \bigcup_{w \in W} P(h')w(V_{\geq 2}(h)).$$

Suppose that $S(h') \subset \overline{S(h)}$. Let $v' \in V_2(h') \cap S(h')$. Then it follows that there are $p \in P(h')$, $w \in W$, $v \in V_{\geq 2}(h)$ with $v' = pw \cdot v$, or $p^{-1} \cdot v' = w \cdot v$.

We have that $P(h') = Z(h) \ltimes N$, where N is the unipotent subgroup of G with Lie algebra $\oplus_{\tau > 0} \mathfrak{g}_\tau(h)$. So $p^{-1} = zn$ with $z \in Z(h)$, $n \in N$. As $v' \in V_2(h')$, we see that $nv' = v' + v''$ with $v'' \in V_{>2}(h')$. So $p^{-1} \cdot v' = zv' + zv''$ with $zv' \in V_2(h')$, $zv'' \in V_{>2}(h')$. In particular, $p^{-1} \cdot v' \in V_{\geq 2}(h')$. But $w \cdot v \in V_{\geq 2}(wh)$. So $p^{-1} \cdot v' \in V_{\geq 2}(h') \cap V_{\geq 2}(wh)$. Denote the latter space by $\widetilde{U}$.

Since h' and wh commute, $\widetilde{U}$ is stable under h'. So $\widetilde{U}$ is the direct sum of h'-eigenspaces. Hence $zv' \in \widetilde{U}$. So, in fact, $zv' \in U$, and obviously, $zv' \in S(h')$.

The last statement follows from [**VP89**, Theorem 5.4]. □

These lemmas underpin a direct method for checking whether $S(h') \subset \overline{S(h)}$:

(1) For all $w \in W$ compute the space $U_w = V_2(h') \cap V_{\geq 2}(wh)$.
(2) Take a random point $u \in U_w$. If $\dim \mathfrak{z}(h') \cdot u = \dim V_2(h')$, then conclude that $S(h') \subset \overline{S(h)}$.

If in Step 2, the equality does not hold, then it is very likely that U_w contains no point of $S(h')$. However, we still need to prove it. One method for that is described in [**GVY12**, Section 5], based on computing the generic rank of a matrix with polynomial entries. It also works here. However, a different approach is also possible: Compute the weights $\mu_1, \ldots, \mu_r \in \mathfrak{h}^*$ of the weight spaces whose sum is U_w. By using the form $(\ ,\ )$ we obtain an isomorphism $\nu : \mathfrak{h} \to \mathfrak{h}^*$ by $\nu(x)(y) = (x, y)$. We consider the Euclidean space $\mathfrak{h}_{\mathbb{R}} = \mathbb{R} \otimes \mathfrak{h}_{\mathbb{Q}}$ with inner product $(\ ,\ )$. Let C be the convex hull in $\mathfrak{h}_{\mathbb{R}}$ of the points $\hat{h}_i = \nu^{-1}(\mu_i)$. Note that all $\hat{h}_i$ lie in the affine space H_2 consisting of all $x \in \mathfrak{h}_{\mathbb{R}}$ with $(h', x) = 2$. So also $C \subset H_2$. Let $\tau \in \mathbb{Q}$ be such that $(h', \tau h') = 2$, then also $\tau h' \in H_2$. Now if $\tau h'$ does not lie in C then U_w has no point of $S(h')$. This follows from the following fact: let $u \in U_w$, and let C' be the convex hull of $\nu^{-1}(\mu)$, where μ ranges over the weights involved in an expression of u as sum of weight vectors, and let $\tilde{h}$ be the point on C' closest to 0, and let $\hat{h}$ be such that $(\tilde{h}, \hat{h}) = 2$, then $\hat{h}$ is a characteristic of u, or $\mathfrak{h}$ does not contain a characteristic of u (see [**VP89**, Section 5.5] or [**Gra17**, Lemma 7.4.16]).

5.3. Implementation for Spin_{13}. The Lie algebra of $G = \mathrm{Spin}_{13}(\mathbb{C})$ is the simple Lie algebra of type B_6. We can construct this Lie algebra in GAP. The nodes of the Dynkin diagram of the root system of this Lie algebra are numbered in the usual way (see, for example, [**Hum78**, §11.4]). Denoting the corresponding fundamental weights by $\lambda_1, \ldots, \lambda_6$ we have that the highest weight of the spinor module is λ_6. The SLA package contains the function `CharacteristicsOfStrata` which implements Popov's algorithm. On input a semisimple Lie algebra and a dominant weight it returns a list of two lists: the first is the list of characteristics, the second is the list of dimensions of the corresponding strata. In the next example we compute the characteristics of the strata of the spinor module of G (which takes about 87 seconds). With `SortParallel` we sort the list of dimensions, and apply the same permutation to the list of characteristics. We display the list of dimensions and the first characteristic, which is an element of $\mathfrak{g}$. Comparing with [**DK15**, Example (f), p. 348] we see that we get the same dimensions as A'Campo and Popov.

```
gap> L:= SimpleLieAlgebra("B",6,Rationals);;
gap> st:= CharacteristicsOfStrata( L, [0,0,0,0,0,1] );; time;
86818
gap> chars:= st[1];; dims:= st[2];;
gap> SortParallel( dims, chars );
gap> dims;
[ 22, 32, 35, 42, 43, 43, 46, 50, 50, 53, 56, 58, 62 ]
gap> chars[1];
(2/3)*v.73+(4/3)*v.74+(2)*v.75+(8/3)*v.76+(10/3)*v.77+(2)*v.78
```

Above we already argued that the strata are G-orbits. In order to be able to apply the algorithm of the previous section we need to show that for each characteristic h the space $V_2(h)$ has an open $Z(h)$-orbit. For this we first construct the spinor module V (this is done with the GAP function `HighestWeightModule`). If `x`, `v` are elements of the Lie algebra `L` and the module `V` respectively, then `x^v` is the result of acting with `x` on `v`. Since the basis elements of the module that is output by `HighestWeightModule` are weight vectors relative to the Cartan subalgebra of `L` that contains the characteristics, the following function can be used to find a basis of $V_2(h)$:

```
V2:= function( V, h )
  return Filtered( Basis(V), v -> h^v = 2*v );
end;
```

Let h be a characteristic, say the fifth one. We show that $V_2(h)$ has an open $Z(h)$-orbit:

```
gap> V:= HighestWeightModule( L, [0,0,0,0,0,1] );;
gap> h:= chars[5];;
gap> v2:= V2( V, h );;
gap> v:= Sum( v2, x -> Random([-100..100])*x );;
gap> zh:= LieCentralizer( L, Subalgebra( L, [h] ) );;
gap> zhv:= Subspace( V, List( Basis(zh), x -> x^v ) );;
gap> Dimension( zhv ) = Length(v2);
true
```

Here we take a random point v of $V_2(h)$. We let `zh`, `zhv` be the centralizer $\mathfrak{z}(h)$ and the space $\mathfrak{z}(h) \cdot v$ respectively. The last line shows that $\dim \mathfrak{z}(h) \cdot v = \dim V_2(h)$. This implies that the orbit of v is open in $V_2(h)$ (Remark 5.1). We have executed this procedure for all characteristics, and hence both hypotheses of the previous section are satisfied.

Now in order to execute the procedure of the previous section we need functions for computing $V_{\geq 2}(h)$ and wh for w in the Weyl group W. The function for the former is straightforward:

```
Vgeq2:= function( V, h )

        local m,i;

        m:= MatrixOfAction( Basis(V), h );
        i:= Filtered( [1..Length(m)], i -> m[i][i] >= 2 );
        return Basis( V ){i};
end;
```

That is, we take the matrix of `h` (which is diagonal) and return the list of basis vectors that correspond to an eigenvalue which is at least 2.

In order to compute wh we consider a Chevalley basis of L, [**Hum78**, Theorem 25.2]. Such a basis consists of elements x_α for α in the root system, and h_1, $\dots$, h_ℓ that lie in the Cartan subalgebra. We refer to the cited theorem for the multiplication table with respect to this basis. For a root α we set $h_\alpha = [x_\alpha, x_{-\alpha}]$. Then we have $wh_\alpha = h_{w\alpha}$. Furthermore, if α_1, $\dots$, α_ℓ are the simple roots then $h_{\alpha_i} = h_i$.

A simple Lie algebra in GAP, constructed with the function `SimpleLieAlgebra`, has a stored Chevalley basis. This is a list consisting of three lists. In the first list we have the x_α for α a positive root. In the second list we have the x_α for α a negative root. The third list has the elements $h_1, \ldots, h_\ell$. The ordering that is used on the positive roots is height compatible (cf. [**Hum78**, §10.1]). This means that the x_{α_i} for $1 \leq i \leq \ell$ come first. Denote the positive roots, as ordered by GAP, by $\alpha_1, \ldots, \alpha_n$. For $n+1 \leq i \leq 2n$ set $\alpha_i = -\alpha_{i-n}$. The SLA package has a function, `WeylGroupAsPermGroup`, that gives the Weyl group as a permutation group on 1, …, $2n$. If w is an element of this group then the corresponding element of the Weyl group acts as $\alpha_i \mapsto \alpha_{i^w}$. These considerations yield the following function for computing wh, where w is given as a permutation and h lies in the given Cartan subalgebra. Here the first two input parameters are the following: `BH` is the basis of the Cartan subalgebra with basis vectors $h_1, \ldots, h_l$; `hs` is the list h_{α_i} for $1 \leq i \leq 2n$.

```
wh:= function( BH, hs, w, h )

         local cf, i;

         cf:= Coefficients( BH, h );
         i:= List( [1..Length(cf)], j -> j^w );
         return cf*hs{i};
end;
```

With this preparation we can give the implementation of the algorithm described in Section 5.2. Here we give the simplified probabilistic version, where we do not prove the non-inclusions. (The complete version is longer as it includes an implemtation of a function to check membership of a convex hull. It has been used to prove the correctness of the diagram in Figure 1, and is available from the author upon request.) We start by defining a number of global variables that will be accessed by the function. Most of these have been explained above. The list `eW` contains all elements of the Weyl group. The function `inc` is a straightforward implementation of the algorithm given in Section 5.2.

```
L:= SimpleLieAlgebra("B",6,Rationals);
st:= CharacteristicsOfStrata( L, [0,0,0,0,0,1] );
chars:= st[1] ; dims:= st[2];;
SortParallel( dims, chars );

V:= HighestWeightModule(L,[0,0,0,0,0,1]);
R:= RootSystem(L);
ch:= ChevalleyBasis(L);
hs:= List( [1..36], i -> ch[1][i]*ch[2][i] );
hs:= Concatenation( hs, -hs );
h:= ch[3];
BH:= Basis( CartanSubalgebra(L), h );
eW:= Elements( WeylGroupAsPermGroup(R) );

inc:= function( h1, h2 )

        local v2, zh1, w, vgeq2, U, u;

        v2:= Subspace( V, V2( V, h1 ) );
        zh1:= BasisVectors( Basis( LieCentralizer( L, Subalgebra( L, [h1]))));
```

```
        for w in eW do
            vgeq2:= Subspace( V, Vgeq2( V, wh( BH, hs, w, h2 ) ) );
            U:= Intersection( v2, vgeq2 );
            if Dimension(U) > 0 then
               u:= Sum( Basis(U), x -> Random([-30..30])*x );
               if Subspace( V, List( zh1, x -> x^u)) = v2 then
                  return true;
               fi;
            fi;
        od;

        return false;

end;
```

We now give a short example of the useage of this function.

```
gap> inc( chars[6], chars[9] ); time;
true
228
gap> inc( chars[6], chars[7] ); time;
false
3923358
```

Here we see that the orbit with the sixth characteristic is contained in the closure of the orbit with the ninth characteristic, but not in the closure of the orbit with the seventh characteristic. The first computation takes 0.2 seconds whereas the second computation takes 3923.3 seconds. This is explained by the fact that for the second computation the entire Weyl group is transversed, which has 46080 elements, whereas the first computation is decided after considering just one element of the Weyl group.

5.4. Closure diagram and stabilizers. By applying the implementation of the previous section we arrive at the Hasse diagram in Figure 1 that displays the closure relation of the orbits in the null cone.

Using Lemma 5.2 it is straightforward to find representatives of the orbits in the null cone. We illustrate this by an example:

```
gap> h:=chars[5];;
gap> v2:=V2( V, h );;
gap> zh:= Basis( LieCentralizer( L, Subalgebra(L,[h])) );;
gap> Length(v2);
32
gap> v:= v2[1]+v2[32];;
gap> Dimension( Subspace( V, List( zh, x -> x^v ) ) );
32
```

This computation shows that the constructed element v is a representative of the orbit corresponding to the fifth characteristic. (We have found it by systematically trying sums of elements of v2; here we do not go into that.)

Given an element $v \in V$ we can consider its stabilizer in $\mathfrak{g}$:

$$\mathfrak{g}_v = \{x \in \mathfrak{g} \mid x \cdot v = 0\},$$

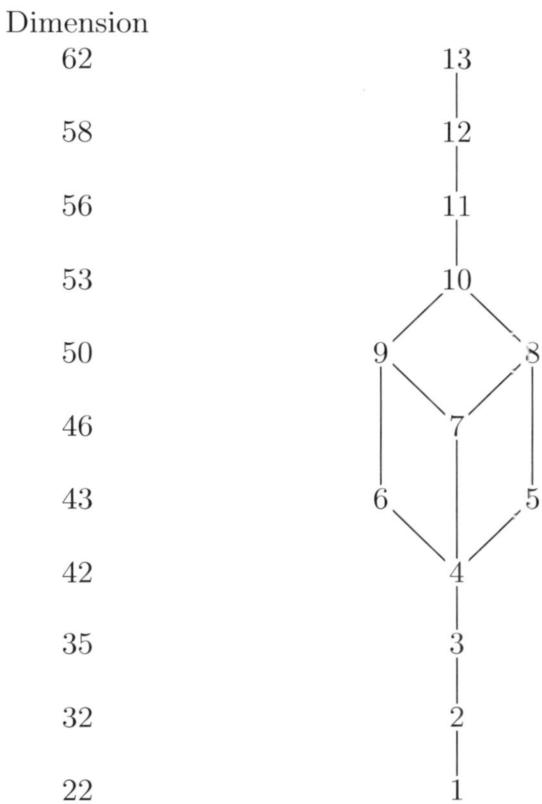

FIGURE 1. Hasse diagram of the closures of the orbits of Spin_{13} in the null cone

which is the Lie algebra of the stabilizer in G. The SLA package does not contain a function for computing this stabilizer, but it is easily written:

```
stab:= function( v ) # v in V, we return its stabilizer in L

       local m, sol;

       m:= List( Basis(L), x -> Coefficients( Basis(V), x^v ) );
       sol:= NullspaceMat( m );
       return List( sol, x -> x*Basis(L) );
end;
```

We then can use GAP functionality to study the structure of the stabilizer. We use the function `LeviMalcevDecomposition` which for a Lie algebra `K` returns a list of two subalgebras. The first of these is semisimple, the second is solvable and `K` is their semidirect sum. In our example this goes as follows.

```
gap> K:= Subalgebra( L, stab(v) );;
gap> ld:=LeviMalcevDecomposition(K);;
gap> SemiSimpleType(ld[1]);
"A4"
```

```
gap> Dimension(ld[2]);
11
```

We see that the stabilizer is the semidirect product of a simple Lie algebra of type A_4 and an 11-dimensional solvable ideal. By inspecting the basis elements of this ideal it is easily seen that it is spanned by root vectors corresponding to positive roots. Hence the ideal is unipotent. We indicate this by saying that the stabilizer is of type $A_4 \ltimes U_{11}$. By doing similar computations for all 13 characteristics we arrive at Table 3.

TABLE 3. Stabilizers of the orbits in the null cone of the spinor representation of Spin_{13}

nr	dim	type of stabilizer
1	22	$A_5 \ltimes U_{21}$
2	32	$A_2 + G_2 \ltimes U_{24}$
3	35	$A_1 + B_3 \ltimes U_{19}$
4	42	$B_2 + T_1 \ltimes U_{25}$
5	43	$A_4 \ltimes U_{11}$
6	43	$C_3 \ltimes U_{14}$
7	46	$B_2 \ltimes U_{22}$
8	50	$A_1 + A_2 \ltimes U_{17}$
9	50	$A_2 \ltimes U_{20}$
10	53	$A_1 + A_1 \ltimes U_{19}$
11	56	$A_1 + A_1 \ltimes U_{16}$
12	58	$B_2 \ltimes U_{10}$
13	62	$A_1 \ltimes U_{13}$

We see that the sum of the dimension in the second column and the dimension of the stabilizer is always $78 = \dim \mathfrak{g}$ (which should be the case as $\dim \mathfrak{g}_v + \dim \mathfrak{g} \cdot v = \dim \mathfrak{g} = 78$).

Acknowledgments

I thank Alexander Elashvili for suggesting the topic of Section 3 and Alexander Premet for suggesting the computations reported on in Section 4. I thank the anonymous referee for many comments which helped to improve the exposition of the paper.

References

[BB92] Philippe Blanc and Jean-Luc Brylinski, *Cyclic homology and the Selberg principle*, J. Funct. Anal. **109** (1992), no. 2, 289–330, DOI 10.1016/0022-1236(92)90020-J. MR1186324

[BJS+19] M. Bocheński, P. Jastrzkebski, A. Szczepkowska, A. Tralle, and A. Woike, *NoCK, nock-package for computing obstruction for compact Clifford-Klein forms., Version 1.4*, `https://pjastr.github.io/NoCK`, Oct 2019, Refereed GAP package.

[BK79] Walter Borho and Hanspeter Kraft, *Über Bahnen und deren Deformationen bei linearen Aktionen reduktiver Gruppen* (German, with English summary), Comment. Math. Helv. **54** (1979), no. 1, 61–104, DOI 10.1007/BF02566256. MR522032

[Bor82] Walter Borho, *Über Schichten halbeinfacher Lie-Algebren* (German, with English summary), Invent. Math. **65** (1981/82), no. 2, 283–317, DOI 10.1007/BF01389016. MR641132

[CdGGT19] S. Cicalò, W. A. de Graaf, and T. GAP Team, *LieRing, computing with finitely presented Lie rings, Version 2.4.1*, `https://gap-packages.github.io/liering/`, Feb 2019, Refereed GAP package.

[CdGSGT19] S. Cicalò, W. A. de Graaf, C. Schneider, and T. GAP Team, *LieAlgDB, a database of Lie algebras, Version 2.2.1*, `https://gap-packages.github.io/liealgdb/`, Oct 2019, Refereed GAP package.

[CM93] David H. Collingwood and William M. McGovern, *Nilpotent orbits in semisimple Lie algebras*, Van Nostrand Reinhold Mathematics Series, Van Nostrand Reinhold Co., New York, 1993. MR1251060

[DFdG20] H. Dietrich, P. Faccin, and W. de Graaf, *CoReLG, computing with real Lie algebras, Version 1.54*, `https://gap-packages.github.io/corelg/`, Jan 2020. Refereed GAP package.

[dGGT19a] W. A. de Graaf and T. GAP Team, *QuaGroup, computations with quantum groups, Version 1.8.2*, `https://gap-packages.github.io/quagroup/`, Oct 2019, Refereed GAP package.

[dGGT19b] W. A. de Graaf and T. GAP Team, *SLA, computing with simple Lie algebras, Version 1.5.3*, `https://gap-packages.github.io/sla/`, Nov 2019, Refereed GAP package.

[DK15] Harm Derksen and Gregor Kemper, *Computational invariant theory*, Second enlarged edition, Encyclopaedia of Mathematical Sciences, vol. 130, Springer, Heidelberg, 2015. With two appendices by Vladimir L. Popov, and an addendum by Norbert A'Campo and Popov; Invariant Theory and Algebraic Transformation Groups, VIII, DOI 10.1007/978-3-662-48422-7. MR3445218

[EG93] Alexander G. Elashvili and Gérard Grélaud, *Classification des éléments nilpotents compacts des algèbres de Lie simples* (French, with English and French summaries), C. R. Acad. Sci. Paris Sér. I Math. **317** (1993), no. 5, 445–447. MR1239028

[GAP21] The GAP Group, *GAP – Groups, Algorithms, and Programming, Version 4.11.1*, 2021.

[GK19] V. Gerdt and V. Kornyak, *FPLSA, finitely presented Lie algebras, Version 1.2.4*, Jan 2019, Refereed GAP package.

[Gra13] Willem A. de Graaf, *Computations with nilpotent orbits in SLA*, `arXiv:1301.1149`, 2013.

[Gra17] Willem Adriaan de Graaf, *Computation with linear algebraic groups*, Monographs and Research Notes in Mathematics, CRC Press, Boca Raton, FL, 2017, DOI 10.1201/9781315120140. MR3675415

[GV78] V. Gatti and E. Viniberghi, *Spinors of 13-dimensional space*, Adv. in Math. **30** (1978), no. 2, 137–155, DOI 10.1016/0001-8708(78)90034-8. MR513846

[GVY12] W. A. de Graaf, E. B. Vinberg, and O. S. Yakimova, *An effective method to compute closure ordering for nilpotent orbits of θ-representations*, J. Algebra **371** (2012), 38–62, DOI 10.1016/j.jalgebra.2012.07.040. MR2975387

[Hes79] Wim H. Hesselink, *Desingularizations of varieties of nullforms*, Invent. Math. **55** (1979), no. 2, 141–163, DOI 10.1007/BF01390087. MR553706

[Hum78] James E. Humphreys, *Introduction to Lie algebras and representation theory*, Graduate Texts in Mathematics, vol. 9, Springer-Verlag, New York-Berlin, 1978. Second printing, revised. MR499562

[Igu70] Jun-ichi Igusa, *A classification of spinors up to dimension twelve*, Amer. J. Math. **92** (1970), 997–1028, DOI 10.2307/2373406. MR277558

[Kac80] V. G. Kac, *Some remarks on nilpotent orbits*, J. Algebra **64** (1980), no. 1, 190–213, DOI 10.1016/0021-8693(80)90141-6. MR575790

[Kra84] Hanspeter Kraft, *Geometrische Methoden in der Invariantentheorie* (German), Aspects of Mathematics, D1, Friedr. Vieweg & Sohn, Braunschweig, 1984, DOI 10.1007/978-3-322-83813-1. MR768181

[LS79] G. Lusztig and N. Spaltenstein, *Induced unipotent classes*, J. London Math. Soc. (2) **19** (1979), no. 1, 41–52, DOI 10.1112/jlms/s2-19.1.41. MR527733

[Pan04] Dmitri I. Panyushev, *On reachable elements and the boundary of nilpotent orbits in simple Lie algebras*, Bull. Sci. Math. **128** (2004), no. 10, 859–870, DOI 10.1016/j.bulsci.2004.08.001. MR2100849

[Pop03] V. L. Popov, *The cone of Hilbert null forms* (Russian, with Russian summary), Tr. Mat. Inst. Steklova **241** (2003), no. Teor. Chisel, Algebra i Algebr. Geom., 192–209; English transl., Proc. Steklov Inst. Math. **2(241)** (2003), 177–194. MR2024052

[PS18] Alexander Premet and David I. Stewart, *Rigid orbits and sheets in reductive Lie algebras over fields of prime characteristic*, J. Inst. Math. Jussieu **17** (2018), no. 3, 583–613, DOI 10.1017/S1474748016000086. MR3789182

[PT14] Alexander Premet and Lewis Topley, *Derived subalgebras of centralisers and finite W-algebras*, Compos. Math. **150** (2014), no. 9, 1485–1548, DOI 10.1112/S0010437X13007823. MR3260140

[SGT18] C. Schneider and T. GAP Team, *Sophus, computing in nilpotent Lie algebras, Version 1.24*, https://gap-packages.github.io/sophus/, Apr 2018, Refereed GAP package.

[VLE18] M. Vaughan-Lee and B. Eick, *LiePRing, database and algorithms for Lie p-rings, Version 1.9.2*, https://gap-packages.github.io/liepring/, Oct 2018, Refereed GAP package.

[VP89] È. B. Vinberg and V. L. Popov, *Invariant theory* (Russian), Algebraic geometry, 4 (Russian), Itogi Nauki i Tekhniki, Akad. Nauk SSSR, Vsesoyuz. Inst. Nauchn. i Tekhn. Inform., Moscow, 1989, pp. 137–314, 315. MR1100485

[Yak10] Oksana Yakimova, *On the derived algebra of a centraliser*, Bull. Sci. Math. **134** (2010), no. 6, 579–587, DOI 10.1016/j.bulsci.2010.03.005. MR2679530

Dipartimento di Matematica, Università di Trento, Italy
Email address: willem.degraaf@unitn.it

Contemporary Mathematics
Volume **783**, 2023
https://doi.org/10.1090/conm/783/15734

Freeness and S-arithmeticity of rational Möbius groups

A. S. Detinko, D. L. Flannery, and A. Hulpke

Abstract. We initiate a new, computational approach to a classical problem: certifying non-freeness of (2-generator, parabolic) Möbius subgroups of $\mathrm{SL}(2,\mathbb{Q})$. The main tools used are algorithms for Zariski dense groups and algorithms to compute a presentation of $\mathrm{SL}(2,R)$ for a localization $R=\mathbb{Z}[\frac{1}{b}]$ of $\mathbb{Z}$. We prove that a Möbius group $G\leq \mathrm{SL}(2,R)$ is not free by showing that it has finite index in $\mathrm{SL}(2,R)$. Further information about the structure of G is obtained; for example, we compute the minimal subgroup of finite index in $\mathrm{SL}(2,R)$ containing G.

1. Introduction

For $x\in\mathbb{C}$, define

$$A(x)=\begin{bmatrix}1 & x\\ 0 & 1\end{bmatrix},\qquad B(x)=\begin{bmatrix}1 & 0\\ x & 1\end{bmatrix}.$$

Let $G(x)$ be the subgroup of $\mathrm{SL}(2,\mathbb{C})$ generated by $A(x)$ and $B(x)$, commonly called a *(parabolic) Möbius group*. Testing freeness of Möbius groups is a well-studied problem; see, e.g., [**1, 3, 12, 15, 17, 18**]. Sanov [**23**] proved that $G(2)$ is free, while Brenner [**2**] proved that $G(x)$ is free (of rank 2) for all x such that $|x|\geq 2$. Hence, if x is algebraic and $|\bar{x}|\geq 2$ for some algebraic conjugate $\bar{x}$ of x, then $G(x)$ is free [**18**, p. 1388]. Also, $G(x)$ is free if x is transcendental [**28**, pp. 30–31].

It is unknown whether $G(x)$ is free for any rational $x\in(0,2)$. Overall, testing freeness of matrix groups is difficult. The problem may be undecidable; note that testing freeness of matrix semigroups is undecidable [**4**]. However, we can effectively decide virtual solvability [**10**], and so, by the Tits alternative, can decide in practice whether a given finitely generated linear group contains a non-abelian free subgroup (without producing one).

On the other hand, non-freeness of $G(x)$ has been justified for infinitely many rationals in $(0,2)$, and there is a set of irrational algebraic x that is dense in $(-2,2)$ and for which $G(x)$ is non-free [**1**, p. 528].

We take a different approach to certifying non-freeness of Möbius groups in $\mathrm{SL}(2,\mathbb{Q})$, that applies recently developed methods to compute with Zariski dense matrix groups [**8, 9**]. We now introduce some basic terms. For an integer $b>1$, let R be the localization $\mathbb{Z}[\frac{1}{b}]=\{r/b^i \mid r\in\mathbb{Z},\ i\geq 0\}$ of $\mathbb{Z}$. Denote $\mathrm{SL}(2,R)$ by Γ.

2020 *Mathematics Subject Classification.* Primary 20-04, 20G15, 20H25, 68W30.

The third author's work has been supported in part by NSF Grant DMS-1720146 and Simons Foundation Grant 852063, which are gratefully acknowledged.

If $I \subset R$ is a non-zero proper ideal, then there is a unique integer $t > 1$ coprime to b such that $I = tR$. Let $\varphi_I\colon \Gamma \to \mathrm{SL}(2, R/I)$ be the associated congruence homomorphism. The kernel Γ_I of φ_I is a *principal congruence subgroup* (PCS) of Γ. We say that the *level* of Γ_I is t, and write φ_t, Γ_t for φ_I, Γ_I, respectively. The following notation will also be used: 1_r is the $r \times r$ identity matrix, $\mathbb{Z}_k := \mathbb{Z}/k\mathbb{Z}$, and $\mathbb{F}_p$ is the field of size p.

Let S be the set of reciprocals of the primes dividing b. A finite index subgroup of Γ is said to be *S-arithmetic*. By [**20**, **25**], Γ has the *congruence subgroup property* (CSP): each S-arithmetic subgroup H of Γ contains a PCS. The *level* of H is defined to be the level of the maximal PCS in H (the PCS in H with smallest possible level).

Throughout, unless stated otherwise, $m = \frac{a}{b} \in \mathbb{Q}$ where a is a positive integer coprime to b. (Note that $G(\frac{1}{b}) = \Gamma$.) Since $G(m) \leq \Gamma$ is Zariski dense in SL(2), the intersection of all S-arithmetic subgroups of Γ that contain $G(m)$ is S-arithmetic [**8**, **9**]. This intersection, called the *arithmetic closure* of $G(m)$, is denoted $\mathrm{cl}(G(m))$. We define the level of $G(m)$ to be the level of $\mathrm{cl}(G(m))$.

If $G(m)$ is S-arithmetic then it is not free (see Section 2). Since Γ is finitely presented, one can attempt to prove S-arithmeticity of $G(m)$ by coset enumeration. This necessitates determining a presentation of Γ. Algorithms for that task are developed in Section 4.1. Then in Section 4.2 we report on experiments carried out using our GAP [**14**] implementation of the algorithms. There we demonstrate S-arithmeticity (hence non-freeness) of $G(m)$ for a range of rational $m \in (0, 2)$. Although non-freeness of some such $G(m)$ was already known, our experiments illustrate the connection between arithmeticity and non-freeness of $G(m)$.

Moreover, we provide essential information about the structure and properties of $G(m)$, covering also the case that $G(m)$ is a *thin matrix group* [**24**], i.e., of infinite index in Γ. Specifically, we prove that $G(m)$ has level a^2 (Theorem 3.7). Hence, if $G(m)$ is S-arithmetic, then we can name the maximal PCS in $G(m)$, and so readily test membership of elements of Γ in $G(m)$. The membership testing problem continues to attract attention; see, e.g., [**6**, **11**].

We are not aware of any rational $m \in (0, 2)$ such that $G(m)$ is thin. If there are none, then this would explain the lack of free $G(m)$ for these m. Conversely, can a thin $G(m)$ be non-free? Within $\mathrm{SL}(2, \mathbb{Z})$, the situation is more settled: $G(2)$ is a free subgroup of finite index, whereas if $m \geq 3$ then $G(m) \leq \mathrm{SL}(2, \mathbb{Z})$ is free and thin [**6**, Theorem 3] (indeed, for integers $m > 5$, the normal closure of $G(m)$ in $\mathrm{SL}(2, \mathbb{Z})$ is thin [**19**, p. 31]). Famously, $\mathrm{SL}(2, \mathbb{Z})$ does *not* have the CSP. We remark that $|\Gamma : G(m)| = \infty$ for any integer $m \geq 1$.

This paper is based on preliminary work presented at the ICERM meeting 'Computational Aspects of Discrete Subgroups of Lie Groups' (June 14–18, 2021).

2. Non-freeness criteria for Möbius groups

Set $A = A(m)$, $B = B(m)$, and $G = G(m)$. Each element of G is a word $W_n = A^{\alpha_1} B^{\beta_1} \cdots A^{\alpha_n} B^{\beta_n}$ where $n \geq 1$ and the α_i, β_j are integers, all of which are non-zero except possibly α_1, β_n. If G is not freely generated by A and B, then G is not free [**18**, p. 1394]. Moreover, G is free and freely generated by A, B if and only if $W_n \neq 1_2$ for each W_n with all exponents α_i, β_j non-zero [**22**, p. 158].

Non-freeness testing of G by a number of authors [**1**, **15**, **18**] depends on finding words of special form in G. The following is a criterion of such type.

LEMMA 2.1. *If $N_G(\langle A \rangle)$ is non-cyclic then G is not free.*

PROOF. The normalizer is upper triangular, hence solvable. □

Since $G \leq G(\frac{m}{n})$, non-freeness (respectively, S-arithmeticity) of G implies that of $G(m/n)$. So in searching for $m \in (0, 2)$ such that G is non-free, we can restrict to $m \in (1, 2)$ if we wish.

Detecting non-identity elements of finite order is another way to prove non-freeness [**5**, **12**]. In [**5**], it is shown that $G(r)$ for $r \in \mathbb{Q}$ has a non-identity element of finite order if and only if $\frac{1}{r} \in \mathbb{Z}$. One direction is simple: $\mathrm{SL}(2, \mathbb{Z}) \subseteq G(\frac{1}{b})$, so, e.g., $-1_2 \in G(\frac{1}{b})$. An elementary proof of the converse is given below.

PROPOSITION 2.2. *For $m = \frac{a}{b}$ where $a > 1$, $G = G(m)$ is torsion-free.*

PROOF. If $h \in G$ has prime order, then h is conjugate to an element of $\mathrm{SL}(2, \mathbb{Z})$. The finite order elements of $\mathrm{SL}(2, \mathbb{Z})$ are known by a result of Minkowski [**22**, p. 179]; of these, only -1_2 is possibly in $G \leq \Gamma_a$. Hence $m = \frac{2}{b}$, b odd. But this cannot be. For an element of G looks like

$$\begin{bmatrix} 1 + m^2 f_1(m) & m f_2(m) \\ m f_3(m) & 1 + m^2 f_4(m) \end{bmatrix}$$

where $f_i(m) \in \mathbb{Z}[m]$, $1 \leq i \leq 4$ ([**5**, p. 747]), and after clearing out denominators in the equation $2f_1(m) = -b^2$, we get a contradiction against b odd. □

Our preferred criterion for non-freeness testing follows.

PROPOSITION 2.3. *Suppose that G is S-arithmetic. Then G is not free.*

PROOF. Since Γ has the CSP, $\Gamma_c \leq G$ for some $c \geq 2$. In turn, Γ_c contains $\langle A(cR) \rangle = \{A(cx) \mid x \in R\}$. The latter is isomorphic to the additive group $(cR)^+$ of the ideal $cR \subseteq R$, and $(cR)^+$ is non-cyclic. □

So, our non-freeness test for Möbius subgroups is really a spinoff from attempts to prove S-arithmeticity in Γ.

3. Exploiting Zariski density of Möbius groups

In this section we establish properties of $G = G(m)$ that derive from Zariski density of G and the fact that Γ has the CSP. In particular, we find a generating set for $\mathrm{cl}(G)$.

Denote the set of prime divisors of $k \in \mathbb{Z}$ by $\pi(k)$. Let $\Pi(H)$ be the (finite) set of all primes p modulo which dense $H \leq \Gamma$ does not surject onto $\mathrm{SL}(2, p)$, i.e.,

$$\Pi(H) = \{p \in \mathbb{Z} \mid p \text{ prime}, \gcd(p, b) = 1, \varphi_p(H) \neq \mathrm{SL}(2, p)\}.$$

LEMMA 3.1. $\Pi(G) = \pi(a)$.

PROOF. If $p \mid a$ then $G \leq \Gamma_p$; so $\pi(a) \subseteq \Pi(G)$. If $p \nmid a$ then $\varphi_p(G) = \mathrm{SL}(2, p)$, because $\mathrm{SL}(2, p) = \langle A(k), B(k) \rangle$ for any $k \in \mathbb{F}_p \setminus \{0\}$. □

Let l be the level of G. Results from [**9**] imply that $\Pi(G) \setminus \{2, 3, 5\} = \pi(l) \setminus \{2, 3, 5\}$. We will see that $l = a^2$ (Theorem 3.7); so $\Pi(G) = \pi(a) = \pi(l)$ without exception.

REMARK 3.2. If G surjects onto $\mathrm{SL}(2, p)$ modulo the prime p, then G surjects onto $\mathrm{SL}(2, \mathbb{Z}_{p^k})$ modulo p^k for all $k \geq 1$.

LEMMA 3.3. *Let $a = p^e c$ where p is a prime, $e \geq 1$, and $\gcd(p, bc) = 1$. Then*

$$\Gamma_{p^{2e}} \leq G\Gamma_{p^f}$$

for any $f \geq 0$, but

$$\Gamma_{p^{2e-1}} \not\leq G\Gamma_{p^f}$$

if $f \geq 2e + 1$.

PROOF. The lemma is trivially true if $f \leq 2e$. Suppose that $f > 2e$. Then $\Gamma_{p^f} \leq \Gamma_{p^{2e+1}} \leq \Gamma_{p^{2e}}$.

Since $\gcd(p, bc) = 1$, there is a positive integer i such that $i \cdot \frac{c}{b} \equiv 1 \bmod p^f$, so

$$A(m)^i \equiv A(p^e) \quad \text{and} \quad B(m)^i \equiv B(p^e) \quad \bmod p^f.$$

Hence $\varphi_{p^f}(G) = \varphi_{p^f}(G(p^e))$. We therefore prove the two assertions with $G(p^e)$ in place of G.

Setting $A(p^e) = x$ and $B(p^e) = y$, we have

$$[x, y] = \begin{bmatrix} 1 + p^{2e} + p^{4e} & p^{3e} \\ -p^{3e} & 1 - p^{2e} \end{bmatrix} \equiv \begin{bmatrix} 1 + p^{2e} & 0 \\ 0 & 1 - p^{2e} \end{bmatrix} \quad \bmod p^{2e+1}.$$

It follows that $U := \varphi_{p^{2e}}(G(p^e))$ is abelian, consisting of the p^{2e} distinct images

$$\varphi_{p^{2e}}\left(\begin{bmatrix} 1 & sp^e \\ tp^e & 1 \end{bmatrix}\right)$$

for $0 \leq s, t < p^e$. But then U does not contain

$$\varphi_{p^{2e}}\left(\begin{bmatrix} 1 + p^{2e-1} & p^{2e-1} \\ -p^{2e-1} & 1 - p^{2e-1} \end{bmatrix}\right).$$

Thus $\varphi_{p^{2e}}(\Gamma_{p^{2e-1}}) \not\leq U$.

It remains to prove the first assertion, and this we do by induction on f. For the base step $f = 2e + 1$, note that the images of $[x, y]$, x^{p^e}, y^{p^e} under $\varphi_{p^{2e+1}}$ are

$$\begin{bmatrix} 1 + p^{2e} & 0 \\ 0 & 1 - p^{2e} \end{bmatrix}, \quad \begin{bmatrix} 1 & p^{2e} \\ 0 & 1 \end{bmatrix}, \quad \begin{bmatrix} 1 & 0 \\ p^{2e} & 1 \end{bmatrix}.$$

These generate the elementary abelian group $\varphi_{p^{2e+1}}(\Gamma_{p^{2e}})$, whose elements are of the form $1_2 + p^{2e}u$ where u has entries in $\{0, \ldots, p-1\}$ and $\operatorname{trace}(u) \equiv 0 \bmod p$.

Assume now that the statement is true for $f = k \geq 2e + 1$. Then $\varphi_{p^k}(\Gamma_{p^{k-1}}) \leq \varphi_{p^k}(G(p^e))$. Thus, for each $1_2 + p^{k-1}u \in \varphi_{p^k}(\Gamma_{p^{k-1}})$ where u is a $\{0, \ldots, p-1\}$-matrix with zero trace modulo p, there exist $v \in G(p^e)$ and some w such that $v = 1_2 + p^{k-1}u + p^k w$. Then $v^p \equiv 1_2 + p^k u \bmod p^{k+1}$, which implies that $\varphi_{p^{k+1}}(\Gamma_{p^k}) \leq \varphi_{p^{k+1}}(G(p^e))$. Hence $\Gamma_{p^{2e}} \leq \langle G(p^e), \Gamma_{p^{k+1}} \rangle$ by the inductive hypothesis. □

REMARK 3.4. Lemma 3.3 is valid for $b = 1$.

LEMMA 3.5. *In the notation of Lemma 3.3, $\varphi_{p^{2e}}(G) \cong C_{p^e} \times C_{p^e}$. Hence*

$$|\operatorname{SL}(2, \mathbb{Z}_{p^{2e}}) : \varphi_{p^{2e}}(G)| = p^{4e} - p^{4e-2} = p^e\,|\operatorname{SL}(2, \mathbb{Z}_{p^e})|.$$

PROOF. We observed that $\varphi_{p^{2e}}(G) = \varphi_{p^{2e}}(G(p^e)) \cong C_{p^e} \times C_{p^e}$ in the proof of Lemma 3.3. Also, $|\operatorname{SL}(n, \mathbb{Z}_{p^k})| = p^{(n^2-1)(k-1)}|\operatorname{SL}(n, p)|$. □

We gather together various observations about the structure of $\operatorname{SL}(2, \mathbb{Z}_{p^e})$, p prime, that are needed in the subsequent proof.

LEMMA 3.6.

(i) *Each proper normal subgroup of* $\mathrm{SL}(2,\mathbb{Z}_{p^e})$ *has index divisible by* p.

(ii) *Let* $l > 1$ *be an integer with prime factorization* $l = p_1^{e_1} \cdots p_k^{e_k}$. *Let* K *be a subgroup of* $\mathrm{SL}(2,\mathbb{Z}_l)$ *such that* $\varphi_{p_i^{e_i}}(K) = \mathrm{SL}(2,\mathbb{Z}_{p^{e_i}})$ *for all* i, $1 \le i \le k$. *Then* $K = \mathrm{SL}(2,\mathbb{Z}_l)$.

PROOF.

(i) A counterexample would arise from a proper normal subgroup of the quotient $\mathrm{PSL}(2,p)$. Hence the statement is clear for $p \ge 5$, and it follows for $p \in \{2,3\}$ by inspection.

(ii) We proceed by induction on k, the base step being trivial. Write $l = rp^e$ where p is the largest prime divisor of l and $\gcd(p,r) = 1$. Then $\mathrm{SL}(2,\mathbb{Z}_l) \cong \mathrm{SL}(2,\mathbb{Z}_{p^e}) \times \mathrm{SL}(2,\mathbb{Z}_r)$; so K is a subdirect product of $\varphi_{p^e}(K) = \mathrm{SL}(2,\mathbb{Z}_{p^e})$ and $\varphi_r(K)$. The inductive hypothesis gives $\varphi_r(K) = \mathrm{SL}(2,\mathbb{Z}_r)$. Then (i) forces G to be the full direct product $\mathrm{SL}(2,\mathbb{Z}_{p^e}) \times \mathrm{SL}(2,\mathbb{Z}_r)$. This completes the proof by induction. □

The next result is the main one of this section, and it facilitates our experiments in the final section.

THEOREM 3.7.

(i) $\mathrm{cl}(G) = G\Gamma_{a^2}$ *has level* a^2.

(ii) $|\Gamma : \mathrm{cl}(G)| = a \cdot |\mathrm{SL}(2,\mathbb{Z}_a)|$.

PROOF.

(i) Let l be the level of G. If $p^e > 1$ is the largest power of the prime p dividing a, then $G\Gamma_{p^{2e}}$ has level p^{2e} by Lemma 3.3. Since $H := \mathrm{cl}(G) = G\Gamma_l \le G\Gamma_{p^{2e}}$ and therefore $\Gamma_l \le \Gamma_{p^{2e}}$, we see that l is divisible by p^{2e}. Thus a^2 divides l.

Next we prove $\pi(l) \subseteq \pi(a)$, i.e., $\pi(l) = \pi(a)$. To this end, suppose that $l = rk$ where $r > 1$, $\pi(k) = \pi(a)$, and $\gcd(r,a) = 1$. By Lemma 3.1, Remark 3.2, and Lemma 3.6, $\varphi_r(G) = \mathrm{SL}(2,\mathbb{Z}_r)$. Since $\mathrm{SL}(2,\mathbb{Z}_l) \cong \mathrm{SL}(2,\mathbb{Z}_r) \times \mathrm{SL}(2,\mathbb{Z}_k)$, it follows that $\varphi_l(H) = \varphi_l(G)$ is a subdirect product of $\mathrm{SL}(2,\mathbb{Z}_r)$ and $\varphi_k(H)$. Each proper quotient of $\mathrm{SL}(2,\mathbb{Z}_r)$ has order divisible by a prime in $\pi(r)$, whereas $\varphi_k(H)$ has order divisible only by the primes in $\pi(a)$. Thus $\varphi_l(H) \cong \mathrm{SL}(2,\mathbb{Z}_r) \times \varphi_k(H)$, and as a consequence $\Gamma_k \le H$. But Γ_l is the PCS of least level in H.

We have now proved that $\pi(l) = \pi(a)$. Therefore $\varphi_l(H)$ is a direct product of p-groups $\varphi_{p^f}(H)$ for p ranging over $\pi(a)$. Suppose that $l > a^2$; say p^f divides l where $f > 2e$ and p^e is the largest power of p dividing a. We infer from $\Gamma_{p^{2e}} \le G\Gamma_{p^f}$ that $\varphi_l(\Gamma_{l/p^{f-2e}}) \le \varphi_l(H)$, so $\Gamma_{l/p^{f-2e}} \le H$: contradiction. Hence $l = a^2$.

(ii) This follows from Lemma 3.5. □

COROLLARY 3.8. $\mathrm{cl}(G)/\Gamma_{a^2} \cong C_a \times C_a$.

The proof of Theorem 3.7 is independent of the denominator b, so we have proved additionally that $G(a) \le \mathrm{SL}(2,\mathbb{Z})$ has level a^2. Cf. [**7**, Proposition 1.12]: for $n \ge 3$, the 'elementary group' in $\mathrm{SL}(n,\mathbb{Z})$ generated by all matrices $1_n + me_{ij}$ with $i \ne j$ (e_{ij} has 1 in position (i,j) and zeros elsewhere) contains the PCS of level m^2.

In line with [**27**] (and cf. [**7**, Proposition 1.10]), one can show that Γ_{a^2} is generated by

$$A(am),\; B(am),\; B(am)^x, \text{ where } x = \begin{bmatrix} -1 & 1 \\ 0 & 1 \end{bmatrix}.$$

Also, at least when b is prime, Γ_{a^2} is the normal closure $\langle A(am)\rangle^\Gamma = \langle B(am)\rangle^\Gamma$ of $G(am)$ in Γ (this implies the CSP; see [**20**]). Thus, we get an explicit generating set of $\mathrm{cl}(G)$.

Theorem 3.7 and Corollary 3.8 afford further insights. Lifting up to G from a presentation of $\varphi_{a^2}(G) \cong C_a \times C_a$ by the 'normal generators' technique (see, e.g., [**10**, §3.2]) yields that $G \cap \Gamma_{a^2}$ is the G-normal closure of

$$\langle A(am), B(am), [A(m), B(m)]\rangle.$$

In summary, using the notation $\leq_{\mathrm{f}}$, $<_\infty$ to denote a subgroup of finite or infinite index, respectively:

(1) if G is thin then $G(am) \leq G <_\infty \mathrm{cl}(G) \leq_{\mathrm{f}} \Gamma_a$;
(2) if G is S-arithmetic then $G(am) \leq \Gamma_{a^2} \leq_{\mathrm{f}} G \leq_{\mathrm{f}} \Gamma_a$;
(3) G is S-arithmetic if and only if $\langle A(am)\rangle^\Gamma \leq G$.

When G is S-arithmetic, an extra benefit of Theorem 3.7 is that it enables us to test membership of $g \in \Gamma$ in G. The idea is obvious: $g \in G$ if and only if $\varphi_{a^2}(g) \in \varphi_{a^2}(G)$. More generally, we could get a negative answer to the question of whether g is in G by testing membership of g in $\mathrm{cl}(G)$.

4. Experimentation

4.1. Computing a presentation of $\mathrm{SL}(2, R)$**.** A key requirement for the experiments, which feature Todd–Coxeter coset enumeration, is a presentation of Γ. Our algorithm to compute one is based on the next theorem, due to Ihara [**26**, Corollary 2, p. 80].

THEOREM 4.1. *For any prime* p,

$$\mathrm{SL}(2, \mathbb{Z}[1/p]) \cong \mathrm{SL}(2, \mathbb{Z}) *_{\Gamma_0(p)} \mathrm{SL}(2, \mathbb{Z}),$$

the free product with amalgamation $\Gamma_0(p)$, *where* $\Gamma_0(p)$ *comprises all matrices in* $\mathrm{SL}(2, \mathbb{Z})$ *that are upper triangular modulo* p.

Theorem 4.1 can be expanded iteratively to $\mathrm{SL}(2, \mathbb{Z}[1/b])$ for composite b [**26**, p. 80]. If k is an integer not divisible by p, then $\mathrm{SL}(2, \mathbb{Z}[1/pk])$ is the amalgamated free product of two copies of $\mathrm{SL}(2, \mathbb{Z}[1/k])$, with the amalgamated subgroup as before comprising all mod-p upper triangular matrices.

We apply Theorem 4.1 in standard fashion (see [**26**, pp. 80–81]), taking the presentations

$$\langle s, t \mid s^4 = 1, (st)^3 = s^2\rangle \quad \text{and} \quad \langle x_p, y_p \mid x_p^4 = 1, (x_py_p)^3 = x_p^2\rangle$$

of $\mathrm{SL}(2, \mathbb{Z})$, where

$$s = \begin{bmatrix} 0 & 1 \\ -1 & 0 \end{bmatrix}, \quad t = \begin{bmatrix} 1 & 0 \\ 1 & 1 \end{bmatrix}, \quad x_p = \begin{bmatrix} 0 & \frac{1}{p} \\ -p & 0 \end{bmatrix}, \quad y_p = \begin{bmatrix} 1 & -\frac{1}{p} \\ 0 & 1 \end{bmatrix}.$$

We construct a generating set for $\Gamma_0(p)$ whose elements are Schreier generators for the stabilizer of the subspace $\langle (1,0)^\top \rangle \subseteq \mathbb{F}_p^2$ under the natural action of $\mathrm{SL}(2, \mathbb{Z})$ modulo p. It is easily seen that $|\mathrm{SL}(2, \mathbb{Z}) : \Gamma_0(p)| = p + 1$. Suppose that $\Gamma_0(p)$ is given by a generating set of matrices $w_i = w_i(x, y)$ that are words in $x = x_p$ and $y = y_p$. Utilizing the GAP package ModularGroup [**16**], we express each w_i as a word $w_i'(s, t)$ in s and t. Then $\mathrm{SL}(2, \mathbb{Z}[1/p])$ has presentation

$$\langle s, t, x, y \mid s^4 = 1, (st)^3 = s^2, x^4 = 1, (xy)^3 = x^2,$$
$$w_1(x, y) = w_1'(s, t), \ldots, w_k(x, y) = w_k'(s, t)\rangle$$

for some $k \leq p+2$; of course, this may simplify. However, eliminating redundancy is usually not worthwhile, because it comes at the cost of longer words.

Below are a few examples that illustrate the method.

- Let $p = 5$. Then $\Gamma_0(p)$ is generated by

$$x^{-2}, xyx^{-1}, y^5, y^2xy^{-2}, yx^{-1}y.$$

Actually, the fifth listed generator is redundant. Now

$$x^{-2} = \begin{bmatrix} -1 & 0 \\ 0 & -1 \end{bmatrix}, \quad xyx^{-1} = \begin{bmatrix} 1 & 0 \\ 5 & 1 \end{bmatrix}, \quad y^5 = \begin{bmatrix} 1 & -1 \\ 0 & 1 \end{bmatrix}, \quad y^2xy^{-2} = \begin{bmatrix} 2 & 1 \\ -5 & -2 \end{bmatrix}.$$

The corresponding words w' are $s^{-2}, t^5, (tst)^{-1}, t^{-2}st^2$. Hence

$$\langle s,t,x,y \mid x^4, xyxyxyx^{-2}, s^4, ststs ts^{-2}, x^{-2}s^2, xyx^{-1}t^{-5}, y^5tst, y^2xy^{-2}t^{-2}s^{-1}t^2\rangle$$

is a presentation of $\mathrm{SL}(2,\mathbb{Z}[1/5])$.

- We similarly calculate that $\mathrm{SL}(2,\mathbb{Z}[1/7])$ has presentation

$$\langle s,t,x,y \mid s^4,ststs ts^{-2}, x^4, xyxyxyx^{-2}, x^{-2}s^2, xyx^{-1}t^{-7}, \\ y^3x^{-1}y^{-2}t^{-3}st^2, y^{-3}x^{-1}y^2t^3st^{-2}\rangle.$$

The generating set in this example is redundant.

- The results for $p = 5$ and $p = 7$ combine to give

$$\mathrm{SL}(2,\mathbb{Z}[\tfrac{1}{35}]) = \langle s,t,x,y,c,d \mid s^4,ststs ts^{-2}, x^4, xyxyxyx^{-2}, x^{-2}s^2, xyx^{-1}t^{-5}, \\ y^5tst, y^2xy^{-2}t^{-2}s^{-1}t^2, c^4, cdcdcdc^{-2}, c^{-2}s^2, \\ cdc^{-1}t^{-7}, d^{-3}c^{-1}d^2t^3st^{-2}, d^3c^{-1}d^{-2}t^{-3}st^2\rangle$$

where $x = x_5$, $y = y_5$, $c = x_7$, and $d = y_7$.

We implemented the above approach as a GAP procedure. This accepts an integer $b > 1$, and returns a presentation of $\Gamma = G(1/b)$ on $2+2|\pi(b)|$ generators. In our experience, this procedure completed rapidly for prime $b \approx 2000$. For composite b, the cost is basically just that of the prime factors, since the amalgamation is straightforward. For larger a, we find short word expressions by systematically enumerating all words by increasing length.

4.2. S-arithmeticity of $G(m)$. In this section we justify S-arithmeticity (and thereby non-freeness) of $G(m)$ for a range of rational m between 0 and 2.

First, we express the generators $A = A(m)$ and $B = B(m)$ of $G = G(m)$ as words in the generators of Γ. When $m = a/p$ for a prime p, this is easy: $A(a/p) = A(1/p)^a = y^{-a}$ and $B(a/p) = sy^as^{-1}$. For larger a, we try to find short word expressions by random search. Then we take the subgroup of Γ defined by the words for A and B and carry out Todd–Coxeter coset enumeration (within the GAP package ACE [**13**]). If the enumeration terminates, then rewriting using the modified Todd–Coxeter algorithm of [**21**] finds relators in A and B. Unless a is small, these relators tend to be long; e.g., the shortest relator discovered for $m = 4/11$ is $A^{121}BA^{-11}B^2A^{-121}B^{-1}A^{11}B^{-2}$. Relator length typically increases with the index of G in Γ (governed by a; Theorem 3.7 (ii)). This suggests that searching for short relators with particular patterns (cf. [**1**]) has limited chance of proving non-freeness. Also recall that if G is S-arithmetic, then it contains upper unitriangular elements u and v such that $\langle u, v\rangle$ is non-cyclic. Once we express $uvu^{-1}v^{-1}$ as a word in A and B, we have a non-trivial relator in G.

Stages in the coset enumeration may entail significantly (and unboundedly) more cosets than $|\Gamma : G|$, so that coset enumeration for G in Γ may fail to terminate within reasonable time or available memory, even for small indices. We emphasize that failure to terminate is *not* a proof that G is thin.

If finite, the index $a\,|\,\mathrm{SL}(2,\mathbb{Z}_a)|$ of $G(a/b)$ in Γ is greater than 10^6, 10^7, 10^8, for $a=33$, $a=59$, $a=101$, respectively. With four generators, and ignoring overhead, each row in the coset table requires at least $4\cdot 2\cdot 8 = 64$ bytes of memory. This means that 1GB of memory can store no more than 10^7 cosets. By this reckoning, we expect that coset enumeration becomes difficult when a is between 40 and 50. If the index is finite, then we expect that enumeration succeeds if $a \leq 25$.

Returning to an earlier point: S-arithmeticity of $G(a/b)$ implies S-arithmeticity of $G(a/kb)$ for every integer $k > 0$, so we could restrict experimentation to groups $G(a/p)$ where p is prime. However, the practicality of an attempt to prove S-arithmeticity of $G(a/b)$ is affected by the size of a. Thus, we investigated $G(m)$ with $m = a/b < 2$ where $b = p^k$, $p \leq 23$ prime. For each fixed denominator b, we found an integer $a_{\max}$ such that $G(a/b)$ is S-arithmetic whenever $a \leq a_{\max}$. Table 1 displays output of the experiments.

TABLE 1. Values of $a_{\max}$ and b such that $G(m)$ is S-arithmetic for all $m = \frac{a}{b}$, where $a \leq a_{\max}$

$a_{\max}$	3	7	13	23	37	45	57	5	14	31	41	9	28	39	11	25
b	2	4	8	16	32	64	128	3	9	27	81	5	25	125	7	49

$a_{\max}$	12	23	15	14	14	14
b	11	121	13	17	19	23

We also managed to prove that $G(m)$ is S-arithmetic for the following $m = a/b$ where a exceeds $a_{\max}$ for the given b: $63/64, 65/64, 44/125, 51/125, 57/125, 29/49$.

Unsuccessful enumerations were re-attempted for comparatively small a. Here we tried strategies such as increased memory, simplifying the presentation, or use of intermediate subgroups. None of these helped, leading us to suspect that the maximum numerator values up to 30 or so are likely to be correct.

Our experiments show that many $G(m)$ previously known to be non-free are S-arithmetic (see [**1**, **12**, **18**]). As a single complementary example, we proved that $G(11/19)$ is not free, whereas freeness of $G(11/19)$ was unresolved in [**1**].

GAP code to accompany this section is posted at `https://github.com/hulpke/arithmetic`.

Acknowledgments

We thank Prof. Vladimir Shpilrain for helpful conversations. We also thank Mathematisches Forschungsinstitut Oberwolfach and Centre International de Rencontres Mathématiques, Luminy, for hosting our visits under their Research Fellowship and Research in Pairs programmes.

References

[1] A. F. Beardon, *Pell's equation and two generator free Möbius groups*, Bull. London Math. Soc. **25** (1993), no. 6, 527–532, DOI 10.1112/blms/25.6.527. MR1245077

[2] Joël Lee Brenner, *Quelques groupes libres de matrices* (French), C. R. Acad. Sci. Paris **241** (1955), 1689–1691. MR75952

[3] J. L. Brenner, R. A. MacLeod, and D. D. Olesky, *Non-free groups generated by two 2×2 matrices*, Canadian J. Math. **27** (1975), 237–245, DOI 10.4153/CJM-1975-029-5. MR372042

[4] Julien Cassaigne, Tero Harju, and Juhani Karhumäki, *On the undecidability of freeness of matrix semigroups*, Internat. J. Algebra Comput. **9** (1999), no. 3-4, 295–305, DOI 10.1142/S0218196799000199. Dedicated to the memory of Marcel-Paul Schützenberger. MR1723469

[5] A. Charnow, *A note on torsion free groups generated by pairs of matrices*, Canad. Math. Bull. **17** (1974/75), no. 5, 747–748, DOI 10.4153/CMB-1974-134-4. MR437652

[6] Anastasiia Chorna, Katherine Geller, and Vladimir Shpilrain, *On two-generator subgroups in $SL_2(\mathbb{Z})$, $SL_2(\mathbb{Q})$, and $SL_2(\mathbb{R})$*, J. Algebra **478** (2017), 367–381, DOI 10.1016/j.jalgebra.2017.01.036. MR3621679

[7] A. S. Detinko, D. L. Flannery, and A. Hulpke, *Algorithms for arithmetic groups with the congruence subgroup property*, J. Algebra **421** (2015), 234–259, DOI 10.1016/j.jalgebra.2014.08.027. MR3272380

[8] A. Detinko, D. L. Flannery, and A. Hulpke, *Zariski density and computing in arithmetic groups*, Math. Comp. **87** (2018), no. 310, 967–986, DOI 10.1090/mcom/3236. MR3739225

[9] A. S. Detinko, D. L. Flannery, and A. Hulpke, *Zariski density and computing with S-integral groups*, preprint (2022).

[10] A. S. Detinko, D. L. Flannery, and E. A. O'Brien, *Algorithms for the Tits alternative and related problems*, J. Algebra **344** (2011), 397–406, DOI 10.1016/j.jalgebra.2011.06.036. MR2831949

[11] Henri-Alex Esbelin and Marin Gutan, *Solving the membership problem for parabolic Möbius monoids*, Semigroup Forum **98** (2019), no. 3, 556–570, DOI 10.1007/s00233-019-10013-4. MR3947312

[12] S. Peter Farbman, *Non-free two-generator subgroups of* $\mathrm{SL}_2(\mathbf{Q})$, Publ. Mat. **39** (1995), no. 2, 379–391, DOI 10.5565/PUBLMAT_39295_13. MR1370894

[13] G. Gamble, A. Hulpke, G. Havas, and C. Ramsay, The GAP package ACE (Advanced Coset Enumerator). `https://www.gap-system.org/Packages/ace.html`

[14] The GAP Group, *GAP – Groups, Algorithms, and Programming*, `http://www.gapsystem.org`

[15] M. Gutan, *Diophantine equations and the freeness of Möbius groups*, Applied Mathematics **5** (2014), 1400–1411. `http://dx.doi.org/10.4236/am.2014.510132`

[16] L. L. Junk and G. Weitze-Schmithüsen, The GAP package ModularGroup, `https://github.com/AG-Weitze-Schmithusen/ModularGroup`

[17] S.-h. Kim and T. Koberda, *Non-freeness of groups generated by two parabolic elements with small rational parameters*, `arXiv:1901.06375v4`

[18] R. C. Lyndon and J. L. Ullman, *Groups generated by two parabolic linear fractional transformations*, Canadian J. Math. **21** (1969), 1388–1403, DOI 10.4153/CJM-1969-153-1. MR258975

[19] Jens L. Mennicke, *Finite factor groups of the unimodular group*, Ann. of Math. (2) **81** (1965), 31–37, DOI 10.2307/1970380. MR171856

[20] J. Mennicke, *On Ihara's modular group*, Invent. Math. **4** (1967), 202–228, DOI 10.1007/BF01425756. MR225894

[21] J. Neubüser, *An elementary introduction to coset table methods in computational group theory*, Groups—St. Andrews 1981 (St. Andrews, 1981), London Math. Soc. Lecture Note Ser., vol. 71, Cambridge Univ. Press, Cambridge-New York, 1982, pp. 1–45. MR679153

[22] Morris Newman, *Integral matrices*, Pure and Applied Mathematics, Vol. 45, Academic Press, New York-London, 1972. MR0340283

[23] I. N. Sanov, *A property of a representation of a free group* (Russian), Doklady Akad. Nauk SSSR (N. S.) **57** (1947), 657–659. MR0022557

[24] Peter Sarnak, *Notes on thin matrix groups*, Thin groups and superstrong approximation, Math. Sci. Res. Inst. Publ., vol. 61, Cambridge Univ. Press, Cambridge, 2014, pp. 343–362. MR3220897

[25] Jean-Pierre Serre, *Le problème des groupes de congruence pour* $\mathbf{SL}_2$ (French), Ann. of Math. (2) **92** (1970), 489–527, DOI 10.2307/1970630. MR272790
[26] Jean-Pierre Serre, *Trees*, Springer-Verlag, Berlin-New York, 1980. Translated from the French by John Stillwell. MR607504
[27] B. Sury and T. N. Venkataramana, *Generators for all principal congruence subgroups of* $\mathrm{SL}(n, \mathbf{Z})$ *with* $n \geq 3$, Proc. Amer. Math. Soc. **122** (1994), no. 2, 355–358, DOI 10.2307/2161024. MR1239806
[28] B. A. F. Wehrfritz, *Infinite linear groups. An account of the group-theoretic properties of infinite groups of matrices*, Ergebnisse der Mathematik und ihrer Grenzgebiete, Band 76, Springer-Verlag, New York-Heidelberg, 1973. MR0335656

Department of Computer Science, School of Computing and Engineering, University of Huddersfield, Huddersfield HD13DH, United Kingdom
Email address: detinko.alla@gmail.com

School of Mathematical and Statistical Sciences, University of Galway, Galway H91TK33, Ireland
Email address: dane.flannery@nuigalway.ie

Department of Mathematics, Colorado State University, Fort Collins, Colorado 80523-1874
Email address: Alexander.Hulpke@colostate.edu

Contemporary Mathematics
Volume **783**, 2023
https://doi.org/10.1090/conm/783/15733

Computability Models: Algebraic, Topological and Geometric Algorithms

Jane Gilman

ABSTRACT. The discreteness problem for finitely generated subgroups of $PSL(2,\mathbb{R})$ and $PSL(2,\mathbb{C})$ is a long-standing open problem. In this paper we consider whether or not this problem is decidable by an algorithm. Our main result is that the answer depends upon what model of computation is chosen. Since our discussion involves the disparate topics of computability theory and group theory, we include substantial background material.

1. Introduction

The discreteness problem for finitely generated subgroups of $PSL(2,\mathbb{R})$ and $PSL(2,\mathbb{C})$ is a long-standing open problem. In this paper we consider whether or not this problem is decidable by an algorithm. Our main result is that the answer depends upon what model of computation is chosen. Since our discussion involves the disparate topics of computability theory and group theory, we include substantial background material.

We ask, *Can one implement a geometric algorithm on a computer?* A geometric algorithm is one given by geometric computations in hyperbolic two-space, $\mathbb{H}^2$. The answer is more complicated than one might think. In this paper we address the issue.

Our work is motivated by M. Kapovich's paper *Discreteness is undecidable.* [**17**]. By this abbreviated statement Kapovich means that the set of finitely generated discrete subgroups of $PSL(2,\mathbb{C})$ is not computable in the sense of Blum-Shub-Smale [**1**].

A Blum-Shub-Smale machine, a BSS machine for short, is one model of computation.

Here we discuss results in different computational models. These include *bit computability* and other models, along with results about their complexity, in [**9**–**11**, **13**].

2020 *Mathematics Subject Classification.* Primary 32G15, 51-08, 20-XX.

Key words and phrases. Discreteness, algorithm, computation, models, Fuchsian groups.

Some of this work was carried out while the author was a supported Visiting Fellow at Princeton University and at Rutgers by a grant from the Rutgers Research Council.

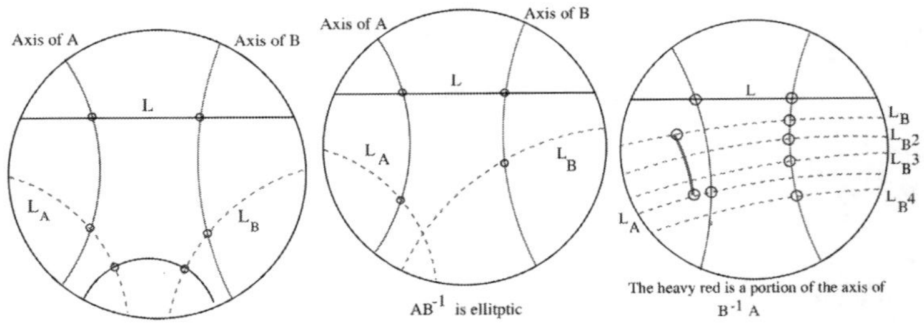

FIGURE 1. Axes and Half-turn Axes Configurations

We consider the various models through the lens of the two-generator $PSL(2,\mathbb{R})$-discreteness question: Namely,

QUESTION 1.1. Given two elements A and B in $PSL(2,\mathbb{R})$ is the group they generate, $G = \langle A, B\rangle$, discrete and non-elementary?

In Theorem 10.1 we show that even though the Gilman-Maskit [**9**] discreteness algorithm establishes decidability of $PSL(2,\mathbb{R})$-discreteness in the geometric model, the BSS model and the symbolic computation model, the algorithm does not work in the bit-computation model and not every step of the GM algorithm is computable in this model.

This yields our main theorem:

THEOREM 1.2. *A problem (resp. a set) which is decidable (resp. computable) in one model of computation is not necessarily decidable (resp. computable) translated to another model of computation.*

The figures above illustrate some configurations that the Gilman-Maskit geometric algorithm might visit as it passes through the steps.

As a specific example, we consider $SL(2,\mathbb{R})$-decidability and we show that for Riemann surfaces of type (g,n), of genus g with n punctures, $T(g,n)$, the Teichmüller Space, and $R(g,n)$ (the rough fundamental domain for the action of the Teichmüller modular group on $T(g,n)$ in the sense of [**18**]) are BSS decidable (Theorem 9.1).

The organization of this paper is as follows: Section 2 contains background of both historical and technical nature and preliminary material. It presents a heuristic definition of an algorithm, some details of the Gilman-Maskit geometric algorithm and an example of the relation of algebra and geometry. Section 3 reviews the definitions of computability and decidability. Section 4 defines what is meant by computation models and these models are discussed in sections 5, 6, 7, 8, touching respectively upon geometry, BSS, symbolic computation and bit-computability. For the most part Sections 6 and 7 contain definitions and summaries of earlier results. The latter are given to place the main theorem in context. The BSS-decidability of $T(g,n)$, Theorem 9.1, is proved in section 9. Section 10 summarizes the results. The last section, Section 11, lists some open questions.

2. Background and Notation

Question 1.1 is an old question, one that mathematicians thought should be easy to solve. As a result there were a number of papers published that had omissions and errors. It turned out that the solution required an algorithm.

2.1. What is an algorithm? Heuristically one can regard an algorithm as a recipe for solving a problem, a recipe that is composed of allowed simple steps and a recipe that always gives the right answer. Since a recipe that does not stop does not give the right answer, the definition implies that an algorithm always comes to a stopping point.[1] However, this definition leaves a lot of room for deciding what the allowed simple steps are. Here we distinguished models of computation in part by consideration of what allowed simple steps for the GM algorithm are included.

2.2. The Gilman-Maskit algorithm. There is a geometric algorithmic solution to the $PSL(2,\mathbb{R})$ discreteness problem due Gilman and Maskit **[9]** .

The Gilman-Maskit algorithm, the GM algorithm for short, consists of easy geometric and computational steps and stops in finite time with an explicit bound.

Pull back to $(\tilde{A},\tilde{B}) \in SL(2,\mathbb{R})$ with tr > 0, det $= 1$ and let T be the maximal initial trace. That is

$T = max\{|\operatorname{tr}\tilde{A}|, |\operatorname{tr}\tilde{B}|, |\operatorname{tr}\tilde{A}\tilde{B}|, |\operatorname{tr}\tilde{A}\tilde{B}^{-1}|\}$

The implementation of the algorithm dovetails tests for Jørgensen 's inequality and hypotheses of the Poincare Polygon Theorem.

NOTATION 2.1. Often in what follows for ease of exposition we will not distinguish notationally between $X \in PSL(2,\mathbb{R})$ and its pull back to $SL(2,\mathbb{R})$. unless it is not clear where the matrix we are referring to sits. We often fail to distinguish between a matrix and its action as a Möbius transformation acting on the hyperbolic plane, $\mathbb{H}^2$.

The algorithm proceeds by considering a given pair of Möbius transformations. There is an order placed on the transformations with hyperbolics harder than parabolics which are in turn harder than elliptics. This induces an order on pairs of transformations. The algorithm considers a pair and either decides that it is not discrete, usually by using Jørgensen's inequality, or determines that is it discrete, usually by using the Poincare Polygon Theorem. If neither discreteness or non-discreteness is determined, the algorithm produces a next pair to consider. The next pair of generators comes from the previous pair using a Nielsen transformation that only changes one of the two generators. The next pair may be in an easier case or it may be in the same type of case but the algorithm assures that the same type of case is repeated at most a finite number of times. Thus the algorithm stops and produces an answer after a finite number of pairs is considered.

2.3. Why emphasize two-generators? A consequence of Jørgensen's inequality is

THEOREM 2.2 (Jørgensen, **[16]**). *Let G be a finitely generated subgroup of $PSL(2,\mathbb{R})$. G is discrete if and only if $\langle A, B\rangle$ is discrete for every pair of $(A,B) \in G$ which generate a non-elementary group.*

This makes two-generator groups especially important.

[1] Riley's **[23]** work produces a *procedure* and not an algorithm because it will not necessarily stop

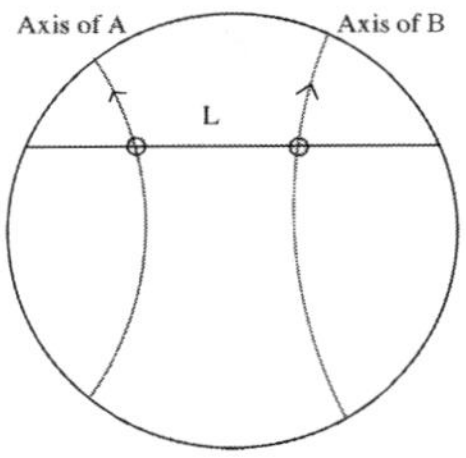

FIGURE 2. Hyperbolics with Coherently Oriented Axes

2.4. Example: Geometric and Algebraic Equivalence. We are interested in translating a problem from one model of computation to another. We give an example of a theorem that translates a geometric condition into an algebra condition and vice-versa. We first note that elements of $PSL(2,\mathbb{C})$ are classified algebraically by their traces and equivalently by the corresponding action on $\mathbb{H}^3$, hyperbolic three-space, geometric conditions. We remind the reader of the definition of coherent orientation. If A and B are hyperbolic transformations with pull backs $\tilde{A}$ and $\tilde{B}$ to $SL(2,\mathbb{R})$ $\operatorname{tr}\tilde{A} \geq \operatorname{tr}\tilde{B} > 2$ with L the common perpendicular to their axes, then A and B are coherently oriented if the attracting fixed points of $\tilde{A}$ and $\tilde{B}$ lie to the left of L when L is oriented from the axis of $\tilde{A}$ towards the axis of $\tilde{B}$ (see Figure 2).[2]

We take the trace of an element $X \in PSL(2,\mathbb{R})$, $\operatorname{tr} X$, to be the trace of the appropriate pull back.

THEOREM 2.3 (Gilman-Maskit [**9**] (1997) page 17). *The Geometric Meaning of Negative Trace*

Assume that $\operatorname{tr} A \geq \operatorname{tr} B > 2$ *and that the axes of* A *and* B *are coherently oriented so that* $\operatorname{tr} AB > 2$. *Then*

(1) $\operatorname{tr} AB^{-1} < -2 \iff$ *the axes of* A, B *and* AB^{-1} *bound a region, and*
(2) $\operatorname{tr} AB^{-1} > 2 \iff$ *one of the axes of* A, B *and* AB^{-1} *separates the other two.*

This is, of course, one of the two main theorems proved in establishing the GM algorithm and showing that it is a true algorithm. The other result is that the procedure replacing a pair of generators by the appropriate Nielsen equivalent pair stops because in the presence of Jørgensen's inequality, there is a positive lower bound by which traces decrease under a Nielsen transformation.

2.5. Overview of the GM Algorithm . The algorithm begins with a pair of elements (A, B) in $PSL(2,\mathbb{R})$ and works with pull backs to $SL(2,\mathbb{R})$ as needed.

The algorithm assumes that beginning pair are a coherently oriented pair of hyperbolics and involves a number of steps to process the pair before it considers a *next* pair of generators or stops. Specifically:

(1) If tr $AB^{-1} \leq -2$,then G is discrete and free.

[2]Let R_X denote reflection in the hyperbolic geodesic X. Given A and B there are hyperbolic geodesics L_A and L_B such that $A = R_{L_A}R_L$ and $B = R_{L_B}R_L$ and $AB^{-1} = R_{L_A}R_{L_B}$. The algorithm proceeds by considering the possible configurations for L, L_A and L_B. The reflection geodesics are pictured in some of the figures but are not needed here.

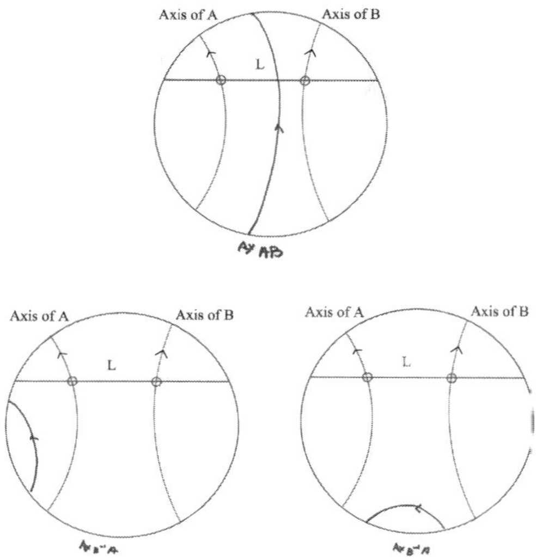

FIGURE 3. Axis BA separates and Axis $B^{-1}A$ bounds

(2) If tr $AB^{-1} > 2$, then repeat 1 and 2 with the coherently oriented pair (AB^{-1}, B) or (B, AB^{-1})
(3) If tr $AB^{-1} = \pm 2$, then go parabolic cases.
(4) If $-2 <$ tr $AB^{-1} < 2$, then G is either not free or not discrete.

For ease of exposition we omit the details of cases that involve parabolics or elliptics. The algorithm can start with a pair that involves parabolics or elliptics. The complexity of all cases is analysed in [**11**] and repeated in the results of theorems 6.3 and 7.1. The complexity of the hyperbolic-hyperbolic case is higher than other cases and thus dominates.

3. Computability and Decidability

The precise definition of a computable function for an oracle Turning machine is given in section 8.

As a starting point we use Braverman's heuristic definition

DEFINITION 3.1 ([**3**, definition 1.11]). A function on a set S is *computable* if there is a Turing machine that takes x as an input and outputs the value $f(x)$.

However, we modify this definition as we vary the model of computation. We will be concerned with functions that are computable in the geometric model, in the BSS model, in the symbolic computation model and in the bit-computability model.

DEFINITION 3.2. A Set S is decidable if and only if there is a computable function with f_S with

$$f_S(x) = \begin{cases} 1, & \iff x \in S; \\ 0, & \iff x \notin S. \end{cases}$$

Sets are decidable (or undecidable) whereas functions are computable (or not).

DEFINITION 3.3. An algorithm is decidable if its stopping set is decidable.

We cite Braverman for a definition of a *Turing Machine*. He says page 1 and 2 of [**4**] "A precise definition of a Turing Machine,TM, is somewhat technical and can be found in all texts on computability e.g . [**21, 24**]." [3][4]

An algorithm will consist of the composition of a number of functions or steps together with branching tests that involve functions. For an algorithm to be computable, the functions that determine steps and branching are required to be computable.

4. Computational models: type of allowed simple steps

We distinguish computation models by the allowed simple steps.

Heuristically a *Turing Machine algorithm* is one where the simple steps can be carried out by a computer. To implement a Turning machine algorithm, a TM algorithm, the input must be finite. Since it may require an infinite amount of information to specify a real number from the set of all real numbers, no TM algorithm can deal with the set of all real numbers. Thus in addition to BSS machines we consider bit-computability and oracle Turing machines (Section 8) that input computable real numbers and symbolic computation 7, where the input consists of rational polynomials or equivalently their coefficients.

We have described the GM algorithm which is a *geometric algorithm.*

Further details of each of the other models can be found in Sections 5, 6, 7, and 8.

5. Geometric Algorithms

A geometric algorithm in hyperbolic two-space is one where the allowed simple steps are geometric in nature. This includes, for example, determining when two geodesics are disjoint or intersect and when given two disjoint geodesics, determining when a third disjoint geodesic separates the other two.

To a mathematician working in hyperbolic geometry, the GM algorithm is a stand alone theorem that determines when a non-elementary two generator group is and is not discrete. The algorithm needs to be translated to other models because the question is often raised as to whether the algorithm has been implemented. It can be implemented using symbolic computation [**11**] but one also needs to think about whether it is bit-computable. This is addressed in Section 8.

6. Blum-Shub-Smale Machines, BSS Machines

We note that the geometric GM algorithm was shown to be decidable in the BSS model [**11**].

We review the details of BSS computation. The reader can also see Braverman [**3**].

[3]He continues, "Such a machine consists of a *tape* and a *head* which can be read/erase/write the symbols on the tape one at a time and can shift its position on the tape in either direction. The symbols on the tape come from a finite alphabet and the TM can be in one of finitely many *states*. Finally a simple look-up table tells TM which action to undertake depending upon the current state and the symbol read on the tape."

[4]We note that there is inconsistent use of terminology-inconsistent use over time partially due to recent developments via computer scientist and bit-computability. Thus we note that what was termed the Real Number Algorithm in [8, 7] would now simple be termed the BSS machine; what was termed the Turning Machine (TM) algorithm there would now be termed the symbolic computation algorithm.

A BSS machine is one where the simple steps

(i) include all ordinary arithmetic operations on real numbers including comparing the size of two real numbers, finding if two real numbers are equal, and determining the sign of a real number.

(ii) include oracles that allow you to compute anything, e.g. computing the arccos of a number, determining the rationality of a real number.

We can consider the steps of a BSS machine as an outline or abstract of the (geometric) algorithm. That is, the sequence of computational moves to be made by the geometric algorithm without taking into account how the input is given or whether the required computational operations can be implemented on a computer.

In particular our BSS machine has oracles so it *can* decide if $\arccos x$ is a rational multiple of π.

Given a BSS machine, the halting is the set S is a set for which given $x \in \mathbb{R}^n$ and an input vector x the BSS machine stops if and only if $x \in \mathbb{R}^n$. If a set S is the halting set of a BSS machine, it is *semicomputable* or *semidecideable*. If its complement is also semidecideable, then the set is BSS computable or BSS decidable.

DEFINITION 6.1. A *semi-algebraic formula* $\phi(x_1, \ldots, x_n)$ is a finite combination of polynomial equalities and inequalities over $\mathbb{R}^n$ linked by the logical connectives of and, or, not. A *semialgebraic set* in $\mathbb{R}^n$ is the set of points satisfying a semi-algebraic formula.

Any semialgebraic set is computable by a BSS machine.

We have the following ([**2**] Theorem 1 and [**3**]).

THEOREM 6.2 (Blum, Shub, Smale [**2**]; Braverman [**3**]). *If a set $C \subset \mathbb{R}^n$ is decided by a BSS machine, then C is a countable union of semi-algebraic sets.*

6.1. Complexities: BSS machine. We define the *algebraic complexity* of an algorithm to be the number of steps it takes to process an input of given size. Let T be **the maximal initial trace**, that is $\max\{|\operatorname{tr} A|, |\operatorname{tr} B|, \operatorname{tr} AB|, |\operatorname{tr} AB^{-1}|\}$ and d the order of the *first finite order elliptic* element the algorithm encounters

We recall

THEOREM 6.3 (Gilman [**11**]). Complexity Part I

(1) *The algebraic complexity of the BSS $PSL(2, \mathbb{Q})$ algorithm is $O(T)$.*
(2) *The algebraic complexity of the BSS $PSL(2, \mathbb{R})$ algorithm is $O(T^2 + d)$*
(3) *The algebraic complexity of one BSS calculation of Jørgensen's inequality is $O(1)$.*

7. Symbolic Computation-Computer Algebra

We note that discreteness was shown to be decidable in the symbolic computation model [**11**]

We assume that the matrix entries lie in a finite simple extension of the rationals $\mathbb{Q}(\gamma)$ of degree D and that the input is given by

M_γ, minimal polynomial for γ; D, the degree; with the eight representing polynomials R_{α_i} each with an isolating interval $Isol_{\alpha_i}$, $i = 1, \ldots, 8$.

We measure size of a polynomial by its degree and its seminorm:

If $P(x) = p_t \cdot x^t + \cdots + p_l \cdot x + p_0$, a polynomial with rational coefficients, $p_i = \frac{r_i}{s_i}$ written in lowest terms.

The **seminorm of** P, $SN(P) = |r_0| + |r_1| + \cdots + |r_t| + |s_0| + \cdots + |s_t|$.
We define $L(SN) = \lfloor \ln(SN) \rfloor + 1$.
$L(SN)$ so behaves computationally very much like a logarithm.

7.1. ∃ Symbolic Computation Algorithms. Given α, β, algebraic numbers, one can do algebraic number arithmetic.

E.g. There are programs [**6, 7**] that input

$$\{R_\alpha, R_\beta, Isol_\alpha, Isol_\beta\}$$

and output

$$\{R_{\alpha+\beta},\ Isol_{\alpha+\beta}\}$$

and also estimate the costs so that for example the cost of addition is $O(D(L(S))^2)$ and that of multiplication is $O(D^3(L(S))^2)$

It is easily seen that the size increases can be estimated by

$$SN(AB) \leq SN(A)SN(B)$$

although better bounds may exist.

7.2. Contrast: GM BSS and Symbolic Computation Translations. We contrast the GM algorithm when translated and implemented as a BSS machine with the algorithm translated and implemented using symbolic computation by comparing the complexities[5].

THEOREM 7.1 (Gilman [**11**]). **Complexity Part II** Theorem 6.3 continued

(4) *The Complexity of* $PSL(2,\mathbb{R})$ *disjoint axes algorithm is*

$$O(2^{k(S_0 D)^2})$$

where k *is a constant.*

S_0 *replaces* T, D *replaces* d *where* D *s the degree of the extension and* S_0 *is the maximum of the semi-norms for the* 8 *representing polynomials and the minimal polynomial of* γ

(5) *The The Complexity of the* first *implementation of Jørgensen is :*

$$O(D^8(L(S_0)^2)$$

But at later step we are looking at S_M *instead of* S_0 *and* $L(S_M)$ *again gives us*

$$O(2^{kS_0^2D^2}).$$

7.3. Exponential growth. For the GM algorithm, the semi-norm, SN, grows exponentially. We note that the algorithm replaces (A,B) by a new pair and the we say that the new pair is either given by a linear step or a Fibonacci step. Specifically replace (A,B) by

$$\begin{cases} (AB^{-1}, B), & \text{linear step;} \\ (B, AB^{-1}), & \text{Fibonacci step.} \end{cases}$$

Note that Fibonacci steps cause the length of the words subsequently considered to grow exponentially and that matrix entries grow exponentially under product. Thus the algorithm is potentially double exponential.

On the other hand, one can handle elliptics

LEMMA 7.2 (Gilman [**11**]). *If* E *is elliptic of finite order, then its order is bound by* $32D^2$

[5]Omitting the intersecting axes case and simplifying notation used in [**11**]

Y.C. Jiang revised parts of the algorithm. He used the revised version to show:

THEOREM 7.3 (Y. C. Jiang [**15**]). *The symbolic computation algorithm is of polynomial complexity.*

This is all well and good except for the fact that as hyperbolic geometers we don't think of all of our points as being roots of polynomials.

8. Bit Computability over the Reals

This is an old topic (see [**19**, **25**, **26**] and reference given there) which has most recently been further developed by Braverman and others [**3**–**5**] since 2005.

We are interested in algorithms over the computable reals which require bit computability. The concept of bit computability can be applied to functions and to graphs and to sets. The model of computation used in these cases is an *oracle Turning machine*.

We begin with definitions(see [**4**]).

DEFINITION 8.1. A function $f(x)$ is computable if there is a TM which takes x as input and outputs the value of $f(x)$.

DEFINITION 8.2. A real number α is said to be computable if there is a computable function $\phi : \mathbb{N} \to \mathbb{Q}$ such that, for all n,

$$|\alpha - \frac{\phi(n)}{2^n}| < 2^{-n}.$$

We work with the set of computable real numbers.

DEFINITION 8.3. A dyadic-valued function $\phi : \mathbb{N} \to \mathbb{D}$ is called an oracle for a real number x if it satisfies $|\phi(m) - x| < 2^{-m}$ for all m.

An *oracle Turing machine* is a TM that can query the value $\phi(m)$ of some oracle for an arbitrary $m \in \mathbb{N}$. We note that in these cases

- Our algorithms cannot say READ x.
- They say instead
  READ x WITH PRECISION $(1/2)^n$.
- The TM comes with this command.

The oracle terminology separates the problem of computing the parameter x from the problem of computing the function f on a given x and note

DEFINITION 8.4. Let $f : S \to \mathbb{R}$. Then $f(x)$ *is computable* if is there is an oracle TM that computes $f(x)$ to any desired degree if accuracy. That is given n, the machines returns a dyadic q with $|q - x| < (1/2)^n$.

DEFINITION 8.5. A set S *is computable* if there is a function that outputs yes if a point is the set and no if the point is not in the set.

THEOREM 8.6 (Braverman, Yampolsky [**4**]). *Let $S \subset \mathbb{R}^k$. Assume that $f : S \to \mathbb{R}^k$ is computable by an oracle TM. Then f is continuous.*

Thus step functions are not computable.[6] That is, $x = 0$, $x = 2$, $>$, $<$ are not computable in this model. The sign function

$$sign(x) = \begin{cases} 1 & \text{if } x \geq 0; \\ 0, & \text{if } x < 0. \end{cases}$$

is not computable.

THEOREM 8.7. *The GM algorithm cannot be translated to a bit computable algorithm and thus it is not decidable in the bit model.*

In [**14**] we address this issue as it arises the attempt translate the GM algorithm to a bit-computable one. We use ideas that include what we term *extended bit-computable domain dependent algorithms* and multi-oracle Turing Machines that do not use a sign oracle. These notions seem appropriate for the algorithms we want to implement as well as for other unrelated algorithms.

9. $SL(2,\mathbb{R})$-decidability: The Teichmüller Space is BSS decidable

The Riemann space or Moduli space is the quotient of the Teichmüller modular group acting on the Teichmüller space of a surface of finite type. A surface of finite type, type (g, n), is a surface of genus g with n punctures where g and n are integers. It is standard to describe $T(g, n)$, the Teichmüller space of a group representing a surface S of finite type by real parameters, trace parameters. The Riemann space can be thought of as the space whose points are conformal equivalence classes surface. The Teichm"̆ller modular group acts on the Teichmüller space yielding Riemann space as its quotient. The space, $R(g, n)$, the rough fundamental domain of [**18**], is an expanded moduli space and is a fundamental domain for the action of the Mapping-class group, but it may contain a finite number of points equivalent to a given point in the Teichmüller space.

In theorem 6.1 of [**18**] the Teichmüller space is given by a series of equalities and inequalities on the trace parameters as is the rough fundamental domain for the action of the Teichm"̆ller modular group.

It follows that

THEOREM 9.1. *Let $S(g, n)$ be a surface of type (g, n), of genus g with n punctures. The Teichmüller space of $S(g, n)$, $T(g, n)$, and $R(g, n)$, the rough fundamental domain for the action of the Teichm"̆ller modular group on $T(g, n)$ are BSS-decidable and are computable sets.*

PROOF. The trace parameters are real parameters for the Teichmüller and Riemann Spaces. The inequalities and equalities in the trace parameters of [**18**] Theorem 6.1 exhibit $T(S)$ and the rough $R(S)$ as semialgebraic sets. By [**1**] semi-algebraic sets are BSS computable. □

[6]Braverman [**5**] has more recent work on how to handle so what he terms *simple functions* and these include step functions. However, these results do not seem relevant to our problem at hand.

10. Conclusion

We have shown

THEOREM 10.1. *The GM algorithm is decidable in the geometric model, the BSS model and the symbolic computation model, it is not decidable in the bit-computable model. Not every step of the algorithm involves a function that is computable in this model.*

PROOF. The GM algorithms requires a step that uses a sign function. □

This proves our main theorem, Theorem 1.2.

11. The Overriding Question

What is the appropriate computational model?
More specifically one can ask

QUESTION 1. Is the question of whether or not an elliptic element of $PSL(2,\mathbb{C})$ is of finite order decidable in some modified bit model?

QUESTION 2. Is the Problem: determine whether $\arccos x$ is a rational multiple of π decidable?

Decidable $\implies$ halting set is semi-algebraic.

Kapovich's theorem, that $PSL(2,\mathbb{C})$ is not decidable in the BSS model depends upon the fact that the Maskit slice is not because cusps which correspond to trace 2 are dense in the boundary. The Maskit slice in the character variety of the punctured torus is the set where $w \in \mathbb{C}$ which has a discrete image for the group generated by the matrices

$$\begin{pmatrix} w & 1 \\ 1 & 0 \end{pmatrix}, \begin{pmatrix} 1 & 2 \\ 0 & 1 \end{pmatrix}.$$

QUESTION 3. Are there interesting subsets of the finitely generated subgroups of $PSL(2,\mathbb{C})$ that are and are not decidable other than the Maskit slice?

QUESTION 4. Is there a way to incorporate $\mathrm{tr} = \pm 2$ into a modified bit-computation model?

12. Acknowledgement

The author thanks the referees for multiple helpful comments.

References

[1] Lenore Blum, Mike Shub, and Steve Smale, *On a theory of computation and complexity over the real numbers: NP-completeness, recursive functions and universal machines*, Bull. Amer. Math. Soc. (N.S.) **21** (1989), no. 1, 1–46, DOI 10.1090/S0273-0979-1989-15750-9. MR974426

[2] Lenore Blum, Felipe Cucker, Michael Shub, and Steve Smale, *Complexity and real computation*, Springer-Verlag, New York, 1998, DOI 10.1007/978-1-4612-0701-6. MR1479636

[3] M. Braverman, *On the Complexity of Real Functions*, preprint, https://arxiv.org/pdf/cs/0502066.pdf

[4] Mark Braverman and Michael Yampolsky, *Computability of Julia sets*, Algorithms and Computation in Mathematics, vol. 23, Springer-Verlag, Berlin, 2009. MR2466298

[5] Mark Braverman and Stephen Cook, *Computing over the reals: foundations for scientific computing*, Notices Amer. Math. Soc. **53** (2006), no. 3, 318–329. MR2208383

[6] G. E. Collins, *Computer algebra of polynomials and rational functions*, Amer. Math. Monthly **80** (1973), 725–755, DOI 10.2307/2318161. MR323750

[7] G. E. Collins and R. Loos, *Real zeros of polynomials*, Computer algebra, Springer, Vienna, 1983, pp. 83–94. MR728967

[8] Werner Fenchel, *Elementary geometry in hyperbolic space*, De Gruyter Studies in Mathematics, vol. 11, Walter de Gruyter & Co., Berlin, 1989. With an editorial by Heinz Bauer, DOI 10.1515/9783110849455. MR1004006

[9] J. Gilman and B. Maskit, *An algorithm for 2-generator Fuchsian groups*, Michigan Math. J. **38** (1991), no. 1, 13–32, DOI 10.1307/mmj/1029004258. MR1091506

[10] Jane Gilman, *Two-generator discrete subgroups of* $\mathrm{PSL}(2,\mathbf{R})$, Mem. Amer. Math. Soc. **117** (1995), no. 561, x+204, DOI 10.1090/memo/0561. MR1290281

[11] Jane Gilman, *Algorithms, complexity and discreteness criteria in* $\mathrm{PSL}(2,\mathbf{C})$, J. Anal. Math. **73** (1997), 91–114, DOI 10.1007/BF02788139. MR1616469

[12] Jane Gilman, *The non-Euclidean Euclidean algorithm*, Adv. Math. **250** (2014), 227–241, DOI 10.1016/j.aim.2013.09.012. MR3122167

[13] Jane Gilman, *Complexity of a Turing machine discreteness algorithm*, In the tradition of Ahlfors and Bers (Stony Brook, NY, 1998), Contemp. Math., vol. 256, Amer. Math. Soc., Providence, RI, 2000, pp. 165–171, DOI 10.1090/conm/256/04004. MR1759677

[14] J. Gilman and A. Tsvietkova, in preparation.

[15] Yicheng Jiang, *Complexity of the Fuchsian group discreteness algorithm*, ProQuest LLC, Ann Arbor, MI, 2000. Thesis (Ph.D.)–Rutgers The State University of New Jersey - Newark. MR2699935

[16] Troels Jørgensen, *A note on subgroups of* $SL(2,\mathbf{C})$, Quart. J. Math. Oxford Ser. (2) **28** (1977), no. 110, 209–211, DOI 10.1093/qmath/28.2.209. MR444839

[17] Michael Kapovich, *Discreteness is undecidable*, Internat. J. Algebra Comput. **26** (2016), no. 3, 467–472, DOI 10.1142/S0218196716500193. MR3506344

[18] Linda Keen, *A rough fundamental domain for Teichmüller spaces*, Bull. Amer. Math. Soc. **83** (1977), no. 6, 1199–1226, DOI 10.1090/S0002-9904-1977-14402-9. MR454075

[19] Ker-I Ko, *Complexity theory of real functions*, Progress in Theoretical Computer Science, Birkhäuser Boston, Inc., Boston, MA, 1991, DOI 10.1007/978-1-4684-6802-1. MR1137517

[20] Curt McMullen, *Cusps are dense*, Ann. of Math. (2) **133** (1991), no. 1, 217–247, DOI 10.2307/2944328. MR1087348

[21] Christos H. Papadimitriou, *Computational complexity*, Addison-Wesley Publishing Company, Reading, MA, 1994. MR1251285

[22] A. Resnick, *Finding the Best Model for Continuous Computation*, Senior Thesis, Harvard College, (2011).

[23] Robert Riley, *Applications of a computer implementation of Poincaré's theorem on fundamental polyhedra*, Math. Comp. **40** (1983), no. 162, 607–632, DOI 10.2307/2007537. MR689477

[24] M. Sipser, *Introduction to the Theory of Computation*, second edition, PWWS Publishing Company, Boston (2005).

[25] Klaus Weihrauch, *Computable analysis*, New computational paradigms, Lecture Notes in Comput. Sci., vol. 3526, Springer, Berlin, 2005, pp. 530–531. MR4376945

[26] Martin Ziegler and Vasco Brattka, *Computability in linear algebra*, Theoret. Comput. Sci. **326** (2004), no. 1-3, 187–211, DOI 10.1016/j.tcs.2004.06.022. MR2094248

Department of Mathematics, Rutgers University, Newark, New Jersey 07102
Email address: gilmanjp@gmail.com

Contemporary Mathematics
Volume **783**, 2023
https://doi.org/10.1090/conm/783/15701

Compact components of planar surface group representations

William M. Goldman

ABSTRACT. Recently Deroin, Tholozan and Toulisse found connected components of relative character varieties of surface group representations in a Hermitian Lie group G with remarkable properties. For example, although the Lie groups are never compact, these components are compact. In this way they behave more like relative character varieties for compact Lie groups. (A relative character variety comprises equivalence classes of homomorphisms of the fundamental group of a surface S, where the holonomy around each boundary component of S is constrained to a fixed conjugacy class in G.)

The first examples were found by Robert Benedetto and myself in an REU in summer 1992. Here S is the 4-holed sphere and $G = \mathsf{SL}(2, \mathbb{R})$. Although computer visualization played an important role in the discovery of these unexpected compact components, computation was invisible in the final proof, and its subsequent extensions.

CONTENTS

2020 *Mathematics Subject Classification.* Primary 57M50; Secondary 22F50, 57K20.

The author acknowledges partial support of the National Science Foundation and the Institute for Advanced Study.

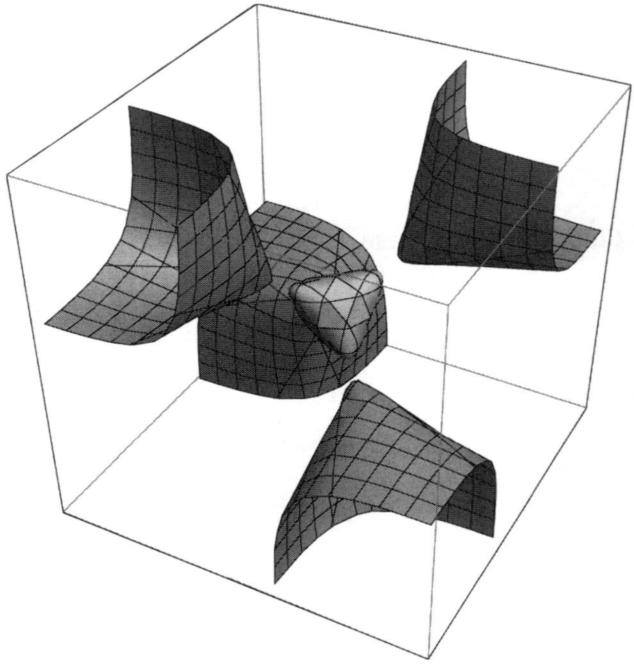

FIGURE 1. A relative character variety over $\mathbb{R}$ with four unbounded components and one compact component.

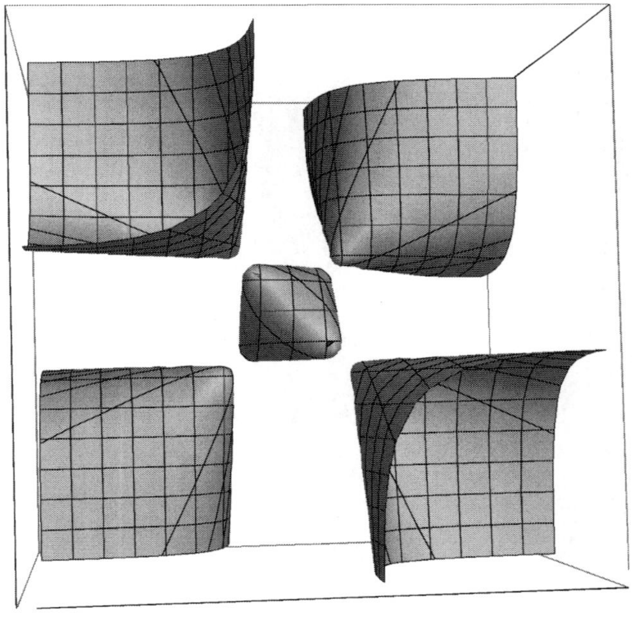

FIGURE 2. Another view of a relative character variety over $\mathbb{R}$ with four unbounded components and one compact component.

1. Relative character varieties of surfaces

Let Σ be a compact oriented surface with boundary $\partial\Sigma = \partial_1 \sqcup \cdots \sqcup \partial_n$ and fundamental group $\pi = \pi_1(\Sigma)$. The *peripheral structure* consists of the conjugacy classes of subgroups $\pi_1(\partial_i) \hookrightarrow \pi$, $i = 1, \ldots, n$ corresponding to the components of $\partial\Sigma$. For a compact orientable surface-with-boundary of genus g and n boundary components we shall write $\Sigma_{g,n}$.

Let G be a reductive linear algebraic group over $\mathbf{k}$ (either $\mathbb{R}$ or $\mathbb{C}$). Then $\mathsf{Hom}(\pi, G)$ is an affine algebraic set with algebraic $\mathsf{Inn}(G)$-action. Let

$$\mathfrak{X}(\Sigma, G) := \mathsf{Hom}(\pi, G)//\mathsf{Inn}(G)$$

be its categorical quotient. Restriction to $\pi_1(\partial_i)$ defines family

$$\mathfrak{X}(\Sigma, G) \longrightarrow \mathfrak{X}(\partial_1, G) \times \cdots \times \mathfrak{X}(\partial_n, G)$$

of *relative character varieties.* The character variety has a natural Poisson structure, for which the relative character varieties are symplectic leaves. The restriction maps are Casimirs for the Poisson structure. In this paper we always work in the classical topology, not the Zariski topology. For background on character varieties, we recommend Sikora [**15**].

In the cases of interest here, this has a very explicit structure, due to the *Vogt-Fricke theorem* which describes the $\mathsf{SL}(2,\mathbb{C})$-character variety of the two-generator free group F_2. Suppose $\mathsf{F}_2 = \langle X, Y\rangle$ be a two-generator free group. Then

$$\mathsf{Hom}\big(\mathsf{F}_2, \mathsf{SL}(2,\mathbb{C})\big) \cong \mathsf{SL}(2,\mathbb{C}) \times \mathsf{SL}(2,\mathbb{C})$$

and $\mathfrak{X}(\mathsf{F}_2, \mathsf{SL}(2,\mathbb{C}))$ is its *Geometric Invariant Theory* quotient under the group $\mathsf{Inn}\big(\mathsf{SL}(2,\mathbb{C})\big) \cong \mathsf{PSL}(2,\mathbb{C})$. The set of $\mathbb{C}$-points of $\mathfrak{X}(\mathsf{F}_2, \mathsf{SL}(2,\mathbb{C}))$ is the quotient space of $\mathsf{Hom}\big(\mathsf{F}_2, \mathsf{SL}(2,\mathbb{C})\big)$ by the equivalence relation where two points are equivalent if and only if their orbit closures contain the same closed[1] orbit. Equivalently it is the *maximal Hausdorff quotient* of $\mathsf{Hom}\big(\mathsf{F}_2, \mathsf{SL}(2,\mathbb{C})\big)$ by $\mathsf{Inn}\big(\mathsf{SL}(2,\mathbb{C})\big)$. See [**7**] for a modern elementary treatment.

Write $Z = (XY)^{-1}$ so that $XYZ = \mathbb{I}$. The $\mathsf{Inn}\big(\mathsf{SL}(2,\mathbb{C})\big)$-invariant mapping

$$\mathsf{Hom}\big(\mathsf{F}_2, \mathsf{SL}(2,\mathbb{C})\big) \xrightarrow{\chi} \mathbb{C}^3$$

$$\rho \longmapsto \begin{bmatrix} x := & \mathsf{Tr}\big(\rho(X)\big) \\ y := & \mathsf{Tr}\big(\rho(Y)\big) \\ z := & \mathsf{Tr}\big(\rho(Z)\big) \end{bmatrix}$$

defines an isomorphism

$$\mathfrak{X}(\mathsf{F}_2, \mathsf{SL}(2,\mathbb{C})) \xrightarrow{\cong} \mathbb{C}^3.$$

This means that every regular function $\mathsf{Hom}\big(\mathsf{F}_2, \mathsf{SL}(2,\mathbb{C})\big) \xrightarrow{f} \mathbb{C}$ which is invariant under $\mathsf{Inn}\big(\mathsf{SL}(2,\mathbb{C})\big)$ factors as $\chi \circ F$ for some polynomial $F \in \mathbb{C}[x, y, z]$. The map χ is constant on closures of $\mathsf{Inn}(\mathsf{SL}(2,\mathbb{C}))$-orbits. A point has a closed orbit if and only if it is either irreducible or is a direct sum of a pair of 1-dimensional representations in $\mathsf{SL}(2,\mathbb{C})$; in the latter case, ρ is equivalent to a representation by

[1] "Closed" refers to the classical topology.

diagonal matrices:

$$\rho(X) = \begin{bmatrix} \xi & 0 \\ 0 & \xi^{-1} \end{bmatrix}$$
$$\rho(Y) = \begin{bmatrix} \eta & 0 \\ 0 & \eta^{-1} \end{bmatrix}$$
$$\rho(Z) = \begin{bmatrix} \zeta & 0 \\ 0 & \zeta^{-1} \end{bmatrix}$$

where $\xi\eta\zeta = 1$, and

$$x = \xi + \xi^{-1}$$
$$y = \eta + \eta^{-1}$$
$$z = \xi\eta + (\xi\eta)^{-1}.$$

An example of such a representation is given in (2.3). The commutator trace

$$\begin{aligned} \mathsf{Hom}\big(\pi, \mathsf{SL}(2,\mathbb{C})\big) &\longrightarrow \mathbb{C} \\ \rho \qquad &\longmapsto \mathsf{Tr}[\rho(X), \rho(Y)] \end{aligned}$$

descends to the polynomial $\kappa \in \mathbb{C}[x, y, z]$ defined by:

$$\kappa(x, y, z) := x^2 + y^2 + z^2 - xyz - 2$$

Then ρ is irreducible if and only if $\kappa(x, y, z) \neq 2$. This condition is equivalent to the $\mathsf{Inn}\big(\mathsf{SL}(2,\mathbb{C})\big)$-orbit being closed and having trivial stabilizer group.[2]

The level set $\kappa(x, y, z) = 2$ corresponds to reducible representations, but the character variety cannot distinguish between a representation and its *semisimplification.* Namely, if ρ is reducible, presrving a linear subspace $L \subset \mathbb{C}^2$, then ρ induces a representation ρ^{ss} on $L \oplus \mathbb{C}^2/L$, which we call its *semisimplification.* Unless ρ is completely reducible (that is, *reductive*) it is *not* conjugate as a representation to ρ^{ss} although it has the same character. The level set $\kappa^{-1}(2)$ is the *Cayley cubic,* discussed in §2 and admits a rational parametrization (2.2). As discussed below, the diagonal matrices are precisely the ones preserving the decomposition $\mathbb{C}^2 \cong L \oplus \mathbb{C}^2/L$; compare (2.3).

The two $\mathbb{R}$-forms of $\mathsf{SL}(2,\mathbb{C})$ are $\mathsf{SU}(2)$ and $\mathsf{SL}(2,\mathbb{R})$. Real characters, that is $(x, y, z) \in \mathbb{R}^3 \subset \mathbb{C}^3$, correspond to equivalence classes of representations into $\mathsf{SU}(2)$ or $\mathsf{SL}(2,\mathbb{R})$. (The common case occurs, namely for representations into the group $\mathsf{SO}(2) = \mathsf{SU}(2) \cap \mathsf{SL}(2,\mathbb{R})$.)

Once again this is detected by the polynomial κ. Suppose that a representation ρ has *real* trace, that is, $(x, y, z) \in \mathbb{R}^3$. If ρ is equivalent to an $\mathsf{SU}(2)$-representation, then $-2 \leq x, y, z \leq 2$. Furthermore, $\rho[X, Y] = [\rho(X), \rho(Y)]$ also has trace in $[-2, 2]$. In particular $\kappa(x, y, z) \leq 2$. Conversely every

$$(x, y, z) \;\in\; [-2, 2]^3 \cap \kappa^{-1}(-\infty, 2]$$

arises as a character of a representation of F_2 into $\mathsf{SU}(2)$.

In [**7**], this is proved by identifying $\mathsf{SU}(2)$ as the universal covering of the orthogonal group $\mathsf{SO}(3)$ which is conjugate to the isometry group of a positive

[2]In other words, the orbit is a closed subset and the action of the group is free.

definite quadratic form on $\mathbb{R}^3$. The corresponding symmetric bilinear form is defined by the symmetric 3×3 matrix

$$\mathbf{B} := \begin{bmatrix} 2 & z & y \\ z & 2 & x \\ y & x & 2 \end{bmatrix}$$

which has determinant $4 - 2\,\kappa(x,y,z)$. The 2×2-minors of $\mathbf{B}$ are positive definite if and only if $-2 < x, y, z < 2$. Furthermore $\mathbf{B}$ itself is positive definite if and only if $-2 < x, y, z < 2$ and $\det(\mathbf{B}) > 0$. Compare [**7**] for further details.

2. The one-holed torus

The fundamental group of $\Sigma_{1,1}$ is a two-generator free group $\langle X, Y\rangle$ with redundant geometric presentation

$$\pi := \langle X, Y, K \mid K = XYX^{-1}Y^{-1}\rangle$$

wit peripheral generator K.

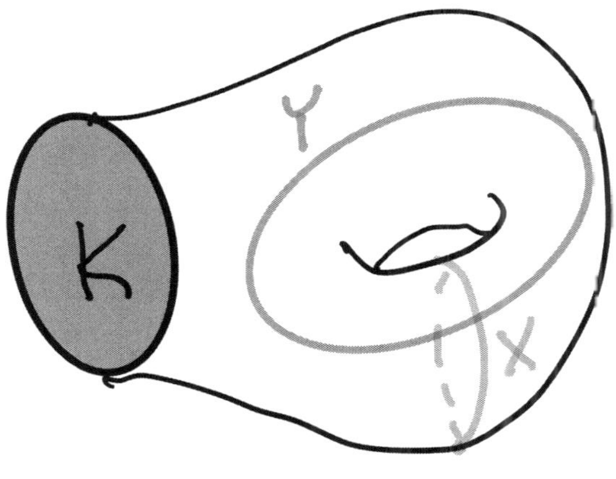

FIGURE 3. Three loops on $\Sigma_{1,1}$.

The boundary trace is defined as follows. The commutator trace function corresponds to the peripheral structure $\partial_1 = K = [X,Y] = XYX^{-1}Y^{-1}$:

$$\begin{aligned} \mathfrak{X}(\mathsf{F}_2, \mathsf{SL}(2,\mathbb{C})) \cong \mathbb{C}^3 &\xrightarrow{\kappa} \mathbb{C} \\ (x,y,z) &\longmapsto x^2 + y^2 + z^2 - xyz - 2 \\ &= \mathsf{Tr}[\rho(X), \rho(Y)] \end{aligned} \tag{2.1}$$

where $x = \mathsf{Tr}(\rho(X)), y = \mathsf{Tr}(\rho(Y)), z = \mathsf{Tr}(\rho(XY))$. Denote the set of $\mathbb{R}$-points of $\kappa^{-1}(k)$ by:

$$\mathcal{S}_k := \kappa^{-1}(k) \cap \mathbb{R}^3$$

Real level sets of κ are relative character varieties:

- For $k < -2$, level set $\mathcal{S}_k$ has four components, each component parametrizing convex hyperbolic structures with totally geodesic boundary whose length l relates to k by:
$$k = -2\cosh(l/2)$$
These convex hyperbolic structures extend uniquely to complete hyperbolic structures with ideal boundary parallel to a unique closed geodesic of length l. Each component is homeomorphic to a disc and the various components are parametrized by spin structures on Σ.
- For $k = -2$, the level set $\mathcal{S}_{-2}$ is the *Markoff surface,* with five components. One component is $\{\mathbf{o}\}$ where $\mathbf{o}$ is the *origin* $(0,0,0)$, the unique $\mathsf{SU}(2)$-character with $\kappa = -2$. It is an isolated point in the real level set $\mathcal{S}_{-2}$, although it is a node in the complexification $\kappa^{-1}(-2)$. The origin is the character of the representation given by the *Pauli spin matrices:*
$$\rho(X) = \begin{bmatrix} 0 & -1 \\ 1 & 0 \end{bmatrix}, \ \rho(Y) = \begin{bmatrix} i & 0 \\ 0 & -i \end{bmatrix}, \ \rho(Z) = \begin{bmatrix} 0 & -i \\ -i & 0 \end{bmatrix}.$$
The other four components correspond to complete finite area hyperbolic structures on $\Sigma_{1,1}$ with spin structures. The *Markoff triples* form the orbit of $(3,3,3)$ under the *mapping class group* $\mathsf{Mod}(\Sigma_{1,1}) \cong \mathsf{GL}(2,\mathbb{Z})$. These correspond to hyperbolic structures on $\Sigma_{1,1}$ with triple symmetry. $\mathbf{o}$ is a singular point in $\kappa^{-1}(-2)$ (an ordinary double point) and is isolated in
$$\mathcal{S}_{-2} = \kappa^{-1}(-2) \cap \mathbb{R}^3.$$
- For $-2 < k < 2$, the level set has five components, one of which is a compact component corresponding to $\mathsf{SU}(2)$-representations (see Figure 5). The four noncompact level sets for $-2 < k < 2$ correspond to hyperbolic structures on torus with an isolated singularity of cone angle θ, where $k = -2\cos(\theta/2)$.
- The level set $\kappa^{-1}(+2)$ corresponds to *reducible characters* and forms the *Cayley cubic*
$$x^2 + y^2 + z^2 - xyz = 4.$$
(Compare Figure 6.) This is the one case when a level set of κ admits a rational parametrization:
$$\begin{gathered} \mathbb{C}^* \times \mathbb{C}^* \longrightarrow \kappa^{-1}(2) \subset \mathbb{C}^3 \\ (\xi,\eta) \longmapsto \begin{bmatrix} \xi + \xi^{-1} \\ \eta + \eta^{-1} \\ \xi\eta + (\xi\eta)^{-1} \end{bmatrix} \end{gathered} \tag{2.2}$$
which is a quotient map by the involution
$$(\xi,\eta) \longmapsto (\xi^{-1},\eta^{-1}).$$
The corresponding reducible representation is:
$$\begin{aligned} X &\xmapsto{\rho} \begin{bmatrix} \xi & * \\ 0 & \xi^{-1} \end{bmatrix} \\ Y &\xmapsto{\rho} \begin{bmatrix} \eta & * \\ 0 & \eta^{-1} \end{bmatrix} \end{aligned} \tag{2.3}$$

- For $k \geq 2$, the level set is homeomorphic to $\Sigma_{0,4}$. In particular it is connected and *noncompact.*

Other famous cubic surfaces (or rather their complex projectivizations) occur in this family: the *Fermat cubic* defined by $A^3+B^3+C^3+D^3=0$ arises for $k=-10/3$ and the *Clebsch diagonal cubic* defined by $A^3+B^3+C^3+D^3+E^3=A+B+C+D+E=0$ arises for $k=18$.

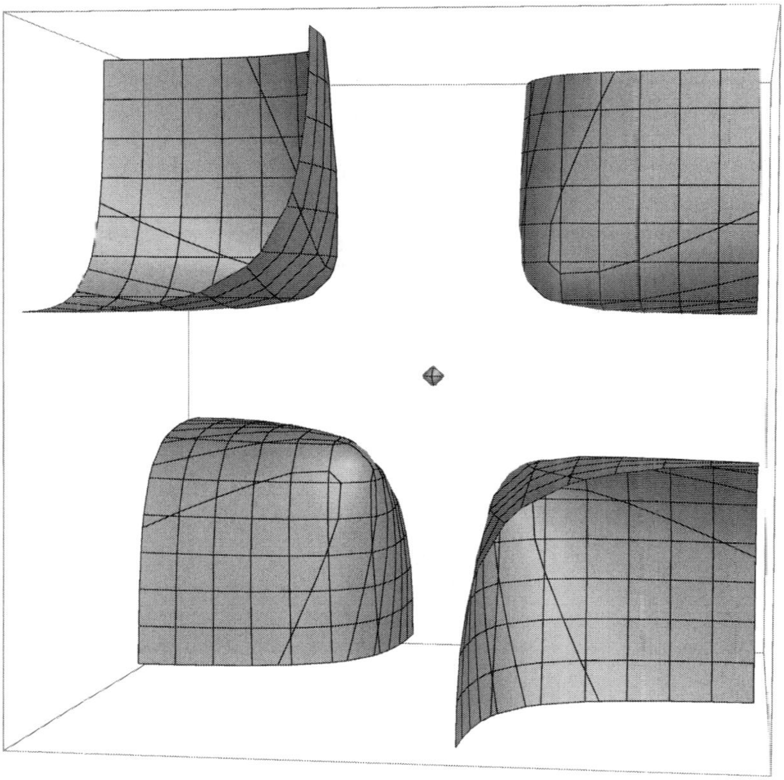

FIGURE 4. The Markoff cubic surface, with the origin and four components corresponding to the Fricke-Teichmüller space of $\Sigma_{1,1}$

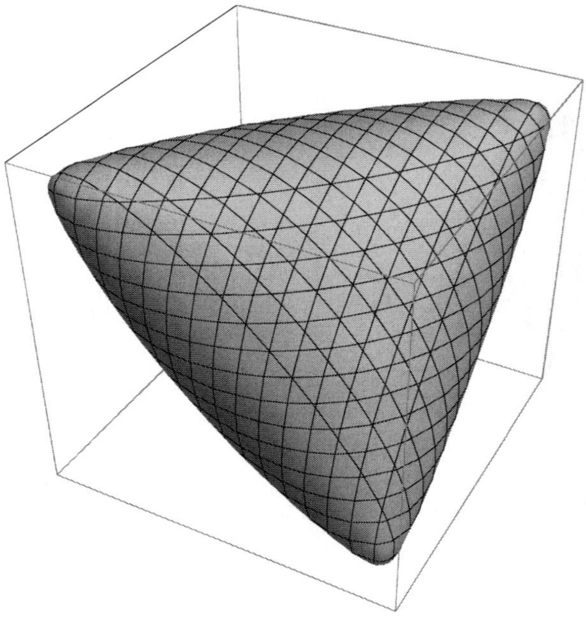

FIGURE 5. A compact component, corresponding to $\mathsf{SU}(2)$-representations of F_2

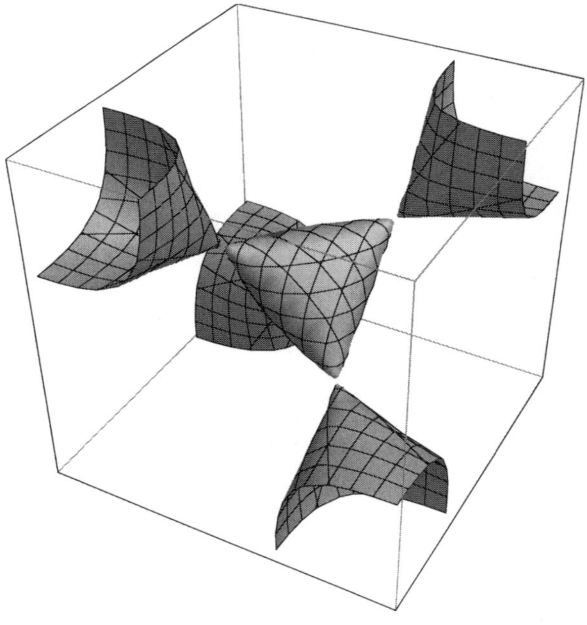

FIGURE 6. The Cayley cubic surface, corresponding to reducible representations. Its four nodes correspond to the (central) $\{\pm\mathbf{1}\}$-representations and form the vertices of a curvilinear tetrahedron.

3. The four-holed sphere

The fundamental group of $\Sigma_{0,4}$ is a three-generator free group $\langle A, B, C\rangle$ given by redundant geometric presentation

$$\pi := \langle A, B, C, D \mid ABCD = \mathbb{I}\rangle$$

with peripheral generators A, B, C, D. For a $\mathsf{SL}(2,\mathbb{C})$-representation ρ, we denote the boundary traces by $a, b, c, d \in \mathbb{C}$. The traces[3] of the interior curves are:

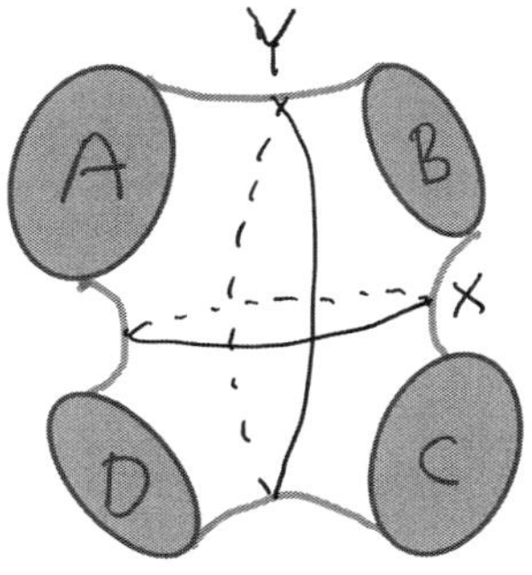

FIGURE 7. Some loops on $\Sigma_{0,4}$.

$$x := -\mathsf{Tr}\big(\rho(AB)\big), \qquad y := -\mathsf{Tr}\big(\rho(BC)\big), \qquad z := -\mathsf{Tr}\big(\rho(CA)\big).$$

These seven functions are related by the defining equation

$$\begin{aligned} x^2 + y^2 + z^2 - xyz + \\ (ab+cd)x + (bc+ad)y + (ca+bd)z \\ = 4 - (a^2+b^2+c^2+d^2+abcd) \end{aligned} \tag{3.1}$$

which we write as

$$x^2 + y^2 + z^2 - xyz + px + qy + rz = s$$

where the linear and constant terms are defined as:

$$p = ab + cd, \tag{3.2}$$

$$q = bc + ad, \tag{3.3}$$

$$r = ca + bd, \tag{3.4}$$

$$s = 4 - (a^2+b^2+c^2+d^2+abcd), \tag{3.5}$$

Then (3.1) defines a quartric hypersurface $\mathcal{V} \subset \mathbb{C}^7$, which we regard as the total space of a family of cubic surfaces with coordinates $(x, y, z) \in \mathbb{C}^3$. This family lives over $\mathbb{C}^4$ with parameters the boundary traces $(a, b, c, d) \in \mathbb{C}^4$. To describe this family of surfaces explicitly, denote the coordinate projection $\mathbb{C}^7 \to \mathbb{C}^3$ by Π_{XYZ}. Then the *relative character varieties*

$$\mathcal{V}_{a,b,c,d} := \Pi_{XYZ}\Big(\mathcal{V} \cap \Big(\{(a,b,c,d)\} \times \mathbb{C}^3\Big)\Big)$$

[3]This convention departs from some previous works ([**1**, **3**, **4**, **7**, **8**]). The minus sign is introduced to make the defining equation (3.1) more compatible with the defining equation (2.1); in particular the higher order terms in both are $x^2+y^2+z^2-xyz$, not $x^2+y^2+z^2+xyz$.

form a family of cubic surfaces in $\mathbb{C}^3$ with coordinates (x, y, z) parametrized by $(a, b, c, d) \in \mathbb{C}^4$. Cantat and Loray [**3**, **4**] prove that the mapping

$$\begin{aligned}\mathbb{C}^4 &\longrightarrow \mathbb{C}^4 \\ (a, b, c, d) &\longmapsto (p, q, r, s)\end{aligned}$$

defined by (3.2),(3.3),(3.4),(3.5) has degree 24, and in particular is surjective. (Compare also Goldman-Toledo [**10**] for a proof of surjectivity.)

When $(a, b, c, d) \in \mathbb{R}^4$ (so $p, q, r \in \mathbb{R}$ as well), the real solutions of (3.1) are either $\mathsf{SL}(2, \mathbb{R})$-characters or $\mathsf{SU}(2)$-characters (or both). In analogy with the case of the one-holed torus, $-2 \leq a, b, c, d \leq 2$ and

$$\kappa(a, b, x), \kappa(c, d, x), \kappa(b, c, y), \kappa(a, d, y), \kappa(c, a, z), \kappa(b, d, z) < 2$$

are *necessary* conditions for a relative character $(a, b, c, d; x, y, z)$ to correspond to an $\mathsf{SU}(2)$-representation. We conjecture that these inequalities are also *sufficient*.

Denote the set of real solutions by

$$\mathcal{S}_{a,b,c,d} := \mathcal{V}_{a,b,c,d} \cap \mathbb{R}^3.$$

In 1992, Benedetto and Goldman [**1**] proved that, for certain $(a, b, c, d) \in [-2, 2]^4$, the real algebraic set $\mathcal{S}_{(a,b,c,d)}$ has a connected component of $\mathsf{SL}(2, \mathbb{R})$-characters which is *compact*. This markedly contrasts the case of relative character varieties of $\Sigma_{1,1}$, when the compact components of $\kappa^{-1}(k) \cap \mathbb{R}^3$ correspond to exactly to $\mathsf{SU}(2)$-representations and not $\mathsf{SL}(2, \mathbb{R})$-representations. (These components exist only if $-2 \leq k < 2$.)

In §9.3 of [**3**], Cantat and Loray show that even if $\mathcal{V}_{a,b,c,d}$ admits a compact component corresponding to $\mathsf{SL}(2, \mathbb{R})$-representations, then by changing (a, b, c, d) but keeping (p, q, r, s) fixed, the compact component corresponds to $\mathsf{SU}(2)$-representations.

The surfaces $\Sigma_{0,4}$ and $\Sigma_{1,1}$ closely relate. When the linear coefficients $p, q, r = 0$, then the defining equation for relative character varieties of $\Sigma_{1,1}$ agrees with that of $\Sigma_{0,4}$ where $k = s - 2$. (Compare [**10**, Theorem 6 and Lemma 7].)

LEMMA 3.1. *Suppose $p, q, r = 0$. Then one of two (not exclusive) possibilities occur:*

- *At least three of a, b, c, d vanish;*
- *Three of a, b, c, d are equal and the fourth equals their negative.*

Both possibilties occur when $a = b = c = d = 0$.

When $a = b = c = d = 0$, we call the character *bi-dihedral*.[4] The corresponding representation sends A, B, C, D to involutions in geodesics in H^3 which admit a common orthogonal geodesic. Such a reprensentation is a double extension of a reducitble representation into $\mathbb{C}^\times < \mathsf{SL}(2, \mathbb{C})$. Here is an example of a bi-dihedral

[4] Apologies for the linguistic impurity.

representation:

$$\begin{aligned} A &\longmapsto \begin{bmatrix} 0 & 1 \\ -1 & 0 \end{bmatrix} \\ B &\longmapsto \begin{bmatrix} 0 & \xi \\ -\xi^{-1} & 0 \end{bmatrix} \\ C &\longmapsto \begin{bmatrix} 0 & \xi\eta \\ -\xi^{-1}\eta^{-1} & 0 \end{bmatrix} \\ D &\longmapsto \begin{bmatrix} 0 & \eta \\ -\eta^{-1} & 0 \end{bmatrix} \end{aligned}$$

It contains the reducible representation (2.3), when the matices are diagonal, with index two.

PROOF OF LEMMA 3.1. Suppose that $d = 0$ but $a, b \neq 0$. Then (3.2) implies that

$$0 \neq ab = ab + cd = p = 0,$$

a contradiction. Thus one of a, b must vanish. If, for example, $a = 0$ and $b, c \neq 0$, then (3.3) implies

$$0 \neq bc = bc + da = q = 0,$$

a contradiction. Thus if one of a, b, c, d vanish, then at least three of them vanish, as claimed.

When $a = b = c = d = 0$, the character is bi-dihedral, as discussed above. Therefore we assume that all $a, b, c, d \neq 0$, and show that three of them are equal and the fourth is their negative.

Definitions (3.3) and (3.4) and $q = r = 0$ imply

$$\frac{a}{b} = -\frac{c}{d}, \qquad \frac{c}{d} = -\frac{b}{a}, \tag{3.6}$$

and $(a/b)^2 = 1$. Thus $a = \pm b$. Suppose first that $a = b$. Then (3.6) implies $d = -c$. Now apply (3.2) with $p = 0$ to similarly conclude that $c = \pm b$. Suppose first that $c = b$. Then $c = b = a = -d$ as desired. Similarly $c = -b$ implies $c = -b = -d = -a$ as desired. The case that $a = -b$ is completely analogous. □

The first case, when the peripheral traces are $(0, 0, 0, d)$, has been treated by §6.3 of Goldman [**6**], and §2.4 of Cantat-Loray [**3**]. In that case the $\Sigma_{0,4}$-character corresponds to a representation ρ where $\rho(A), \rho(B), \rho(C)$ are lifts to $\mathsf{SL}(2, \mathbb{C})$ of involutions in $\mathsf{PSL}(2, \mathbb{C})$ and

$$\rho(D) = \rho(C)\rho(B)\rho(A)$$

is (essentially) their product. The representation thus descends from $\pi_1(\Sigma_{0,4}) \cong \mathsf{F}_3$ to the fundamental group of an orbifold-with-boundary $\mathfrak{O}$ whose underlying surface is a disc and has three branch points of order two. The (hyper-) elliptic involution on $\Sigma_{1,1}$ defines an orbifold-covering space $\Sigma_{1,1} \to \mathfrak{O}$ and the corresponding $\Sigma_{1,1}$-character corresponds to the restriction of the representaion to the induced monomorphism

$$\begin{aligned} \pi_1(\Sigma_{1,1}) &\hookrightarrow \pi_1(\mathfrak{O}) \\ X &\longmapsto A'B' \\ Y &\longmapsto B'C' \end{aligned}$$

where

$$\pi_1(\mathfrak{O}) = \langle A', B', C', D' \mid (A')^2 = (B')^2 = (C')^2 = A'B'C'D' = e\rangle$$

in the notation of [**6**, §6.3], to which we refer for details. A similar interpretation for the relative characters for $(a, a, a, -a)$ for $a \neq 0$ would be interesting.

4. Recent developments

Recently Deroin-Tholozan [**5**] found compact components of $\mathsf{PSL}(2,\mathbb{R})$-characters in $\mathfrak{X}\big(\Sigma_{0,n}, \mathsf{PSL}(2,\mathbb{R})\big)$ for all $n \geq 3$. They called these representations *supra-maximal* since their relative Euler class "exceeded" the presumed maximum value in the Milnor-Wood inequality. Namely, for a surface with nonempty boundary, to define the Toledo invariant (which agrees with the Euler class when $G = \mathsf{PSL}(2,\mathbb{R})$), one requires some boundary conditions. As in [**2**], one must correct the definition of the Toledo invariant when the holonomy around a boundary component c is elliptic. The correction term is the rotation number. Denote the subset of $\mathsf{PSL}(2,\mathbb{R})$ consisting of elliptic elements by Ell, and the subset of hyperbolic elements by Hyp. The *rotation angle mapping* $\mathsf{Ell} \xrightarrow{\theta} (0, 2\pi)$ is $\mathsf{Inn}\big(\mathsf{PSL}(2,\mathbb{R})\big)$-invariant and assigns to

$$\begin{bmatrix} \cos(\theta/2) & -\sin(\theta/2) \\ \sin(\theta/2) & \cos(\theta/2) \end{bmatrix}$$

the parameter θ.[5] Although it is continuous on Ell, it does *not* extend continuously to $\mathsf{PSL}(2,\mathbb{R})$. Deroin and Tholozan extend θ to an invariant upper-semicontinuous function which vanishes on Hyp and takes the identity element $\mathbf{1} \in \mathsf{PSL}(2,\mathbb{R})$ to 2π. ($\theta = 0$ on the positive parabolic elements and $\theta = 2\pi$ on the negative parabolic elements.[6]) Then the corrected relative Euler number $e(\rho)$ of a $\mathsf{PSL}(2,\mathbb{R})$-representation ρ is obtained by lifting the interior generators of $\pi_1(\Sigma)$ to the universal covering $\widetilde{\mathsf{SL}(2,\mathbb{R})}$ and correcting by contributions of θ for each boundary component. Specifically, in terms of the standard presentation

$$\pi_1(\Sigma) = \langle A_1, B_1, \dots, A_g, B_g, C_1, \dots C_n \mid [A_1, B_1] \dots [A_g, B_g] C_1 \dots C_n = \mathbb{I}\rangle$$

and a representation ρ, the expression

$$[\widetilde{\rho(A_1)}, \widetilde{\rho(B_1)}] \dots [\widetilde{\rho(A_g)}, \widetilde{\rho(B_g)}] \quad \widetilde{\rho(C_1)} \dots \widetilde{\rho(C_n)}$$

lies in

$$\pi_1\big(\mathsf{PSL}(2,\mathbb{R})\big) = \mathsf{Ker}\big(\widetilde{\mathsf{SL}(2,\mathbb{R})} \longrightarrow \mathsf{PSL}(2,\mathbb{R})\big) \cong \mathbb{Z},$$

where $\widetilde{X}$ denotes the lift of $X \in \mathsf{PSL}(2,\mathbb{R})$ which is compatible with the above choice of θ. This is the *relative Euler number* $e(\rho)$.

Using the theory of the Toledo invariant and maximal representations developed by Burger-Iozzi-Wienhard [**2**], Deroin and Tholozan find compact components of relative $\mathsf{SL}(2,\mathbb{R})$-characters exist only for *planar surfaces,* that is, when Σ has genus 0. They show that for planar surfaces, their invariant may exceed the Milnor-Wood bound $|\chi(\Sigma)| = n - 2$, but is no greater than n. The case of n only occurs for the trivial representation, and $n - 1$ for the *Deroin-Tholozan* representations.

[5] A. Maret has pointed out that this parameter θ is not completely well-defined until one requires, for example, that $\theta < \pi$ for elliptic elements close to $\mathbf{1}$ in the positive segment of an elliptic one-parameter subgroup.

[6] A positive parabolic element is one $\mathsf{PSL}(2,\mathbb{R})$-conjugate to $\left[\begin{smallmatrix} 1 & 0 \\ t & 1 \end{smallmatrix}\right]$ for $t > 0$ and negative parabolic for $t > 0$. This distinction disappears for $\mathsf{PGL}(2,\mathbb{R})$-conjugacy.

Furthermore they show that if $e(\rho) > n-2$, then the surface is planar ($g = 0$) and compact components arise.

For every choice of boundary traces, each compact component of $\mathfrak{X}(\Sigma_{0,n}, \mathsf{SL}(2,\mathbb{R}))$ is symplectomorphic to $\mathbb{C}\mathsf{P}^{n-3}$ with its standard Fubini-Study symplectic structure, at least up to scale. Using Delzant's theory of moment polytopes, they compute the symplectic volume in terms of the boundary parameters. Arnaud Maret [**13**] found action-angle coordinates for the Hamiltonian twist flows on Deroin-Tholozan components.

Unlike components of Fuchsian characters, $\rho(x)$ is elliptic for every $x \in \pi$ corresponding to a *simple closed curve*. This is easy to see because otherwise the Hamiltonian twist flow would be unbounded, contradicting compactness. They also showed that $\mathsf{Mod}(\Sigma)$-orbit of $[\rho]$ is bounded. Maret [**12**] showed that the the $\mathsf{Mod}(\Sigma)$-action is ergodic with respect to the symplectic measure.

Following suggestions of Olivier Biquard, Gabriele Mondello [**14**] interpreted these results in terms of parabolic Higgs bundles. This is closely related to the fact (see [**5**]: For *every* complex structure on $\Sigma_{0,n}$, $\exists$ ρ-equivariant holomorphic map $\widetilde{\Sigma_{0,n}} \longrightarrow G/K$. This is analogous to constant map when G is compact. Furthermore it contrasts the situation in *higher Teichmüller theory* that for many classes of surface group representations (for example Hitchin representations into low rank simple real forms), that there is a unique conformal structure giving an equivariant holomorphic metric (Labourie [**11**]). This gives a holomorphic identification of the symplectic leaves as above.

Finally we mention that the $\mathsf{PSL}(2,\mathbb{R})$-theory extends to higher rank Lie groups. Tholozan-Toulisse [**16**] have found compact components of representations in higher rank Hermitian Lie groups:

$$\mathsf{U}(p,q), \quad \mathsf{PSp}(2m,\mathbb{R}), \quad \mathsf{SO}^*(2m)$$

and many of the results proved in [**5**] generalize to these groups.

Acknowledgment

I would like to thank Arnaud Maret for various suggestions and corrections on this manuscript.

References

[1] Robert L. Benedetto and William M. Goldman, *The topology of the relative character varieties of a quadruply-punctured sphere*, Experiment. Math. **8** (1999), no. 1, 85–103. MR1685040

[2] Marc Burger, Alessandra Iozzi, and Anna Wienhard, *Surface group representations with maximal Toledo invariant*, Ann. of Math. (2) **172** (2010), no. 1, 517–566, DOI 10.4007/annals.2010.172.517. MR2680425

[3] S. Cantat and F. Loray, *Holomorphic dynamics, Painlevé VI equation and character varieties*, (arXiv:0711.1582v2)

[4] Serge Cantat and Frank Loray, *Dynamics on character varieties and Malgrange irreducibility of Painlevé VI equation* (English, with English and French summaries), Ann. Inst. Fourier (Grenoble) **59** (2009), no. 7, 2927–2978. MR2649343

[5] Bertrand Deroin and Nicolas Tholozan, *Supra-maximal representations from fundamental groups of punctured spheres to* PSL(2, ℝ) (English, with English and French summaries), Ann. Sci. Éc. Norm. Supér. (4) **52** (2019), no. 5, 1305–1329, DOI 10.24033/asens.2410. MR4057784

[6] William M. Goldman, *Ergodic theory on moduli spaces*, Ann. of Math. (2) **146** (1997), no. 3, 475–507, DOI 10.2307/2952454. MR1491446

[7] William M. Goldman, *Trace coordinates on Fricke spaces of some simple hyperbolic surfaces*, Handbook of Teichmüller theory. Vol. II, IRMA Lect. Math. Theor. Phys., vol. 13, Eur. Math. Soc., Zürich, 2009, pp. 611–684, DOI 10.4171/055-1/16. MR2497777

[8] William M. Goldman, *The modular group action on real* SL(2)*-characters of a one-holed torus*, Geom. Topol. **7** (2003), 443–486, DOI 10.2140/gt.2003.7.443. MR2026539

[9] W. Goldman, *Geometric Structures on Manifolds,*, Graduate Studies in Mathematics, vol. 227, American Mathematical Society, Providence, RI, 2023.

[10] W. Goldman and D. Toledo, *Affine cubic surfaces and relative* SL(2)*-character varieties of compact surfaces,* arXiv:1006.3838v2 (unpublished)

[11] François Labourie, *Cyclic surfaces and Hitchin components in rank 2*, Ann. of Math. (2) **185** (2017), no. 1, 1–58, DOI 10.4007/annals.2017.185.1.1. MR3583351

[12] A. Maret, *Ergodicity of the mapping class group action on super-maximal representations,* arXiv:2012.05775

[13] A. Maret, *Action-angle coordinates for surface group representations in genus zero,* arXiv:2110.13896

[14] Gabriele Mondello, *Topology of representation spaces of surface groups in* $\mathrm{PSL}_2(\mathbb{R})$ *with assigned boundary monodromy and nonzero Euler number*, Pure Appl. Math. Q. **12** (2016), no. 3, 399–462, DOI 10.4310/PAMQ.2016.v12.n3.a3. MR3767231

[15] Adam S. Sikora, *Character varieties*, Trans. Amer. Math. Soc. **364** (2012), no. 10, 5173–5208, DOI 10.1090/S0002-9947-2012-05448-1. MR2931326

[16] Nicolas Tholozan and Jérémy Toulisse, *Compact connected components in relative character varieties of punctured spheres* (English, with English and French summaries), Épijournal Géom. Algébrique **5** (2021), Art. 6, 37, DOI 10.46298/epiga.2021.volume5.5894. MR4275255

Department of Mathematics, University of Maryland, College Park, Maryland 20742

Current address: School of Mathematics, Institute for Advanced Study, Princeton, New Jersey 08540

Email address: wmg@umd.edu

Contemporary Mathematics
Volume **783**, 2023
https://doi.org/10.1090/conm/783/15699

Proving infinite index for a subgroup of matrices

Alexander Hulpke

ABSTRACT. We show how tools from computational group theory can be used to prove that a subgroup of matrices has infinite index.

The combination of algorithmic methods for matrix groups [**1**] with those for finitely presented groups [**8**] has been used successfully [**4**] to prove that certain subgroups of infinite matrix groups have finite index. The purpose of this note is to show, in a concrete example, how the same toolkit can be used to prove the infinite index of a particular subgroup. All calculations were performed in GAP [**5**], a transcript of the calculation and code is available at `https://www.math.colostate.edu/~hulpke/paper/MaxIndexTranscript.txt`.

The group in question in this example is $\Gamma = \mathrm{GL}_2(\mathbb{Z}(\zeta))$, for ζ a primitive 3rd root of unity. Motivated by [**2**], A. Bächle asked (private communication), whether the subgroup $S = \langle m_1, m_2, m_3, m_i, m_j, m_t\rangle$, with

$$m_1 = \zeta\begin{pmatrix} 0 & -112 \\ 112 & 0 \end{pmatrix} + \zeta^2\begin{pmatrix} 97 & -56 \\ 56 & 97 \end{pmatrix}, \qquad m_i = \begin{pmatrix} 0 & 1 \\ -1 & 0 \end{pmatrix},$$

$$m_2 = 56\zeta\begin{pmatrix} 1 & 1 \\ 1 & -1 \end{pmatrix} + \zeta^2\begin{pmatrix} 41 & 112 \\ 112 & 153 \end{pmatrix}, \qquad m_j = \zeta\begin{pmatrix} 1 & \zeta \\ \zeta & -1 \end{pmatrix},$$

$$m_3 = 56\zeta\begin{pmatrix} 1 & -1 \\ -1 & -1 \end{pmatrix} + \zeta^2\begin{pmatrix} 209 & 56 \\ 56 & -15 \end{pmatrix}, \qquad m_t = \zeta\begin{pmatrix} 1 & 0 \\ 0 & 1 \end{pmatrix}$$

has finite index in Γ.

We obtain a finite presentation for Γ from [**9**, Theorem 6.1], which gives us that $\Gamma \cong \langle t, u, j, l, a, w\rangle$ with

$$t = \begin{pmatrix} 1 & 1 \\ 0 & 1 \end{pmatrix}, \qquad u = \begin{pmatrix} 1 & \zeta \\ 0 & 1 \end{pmatrix}, \qquad j = \begin{pmatrix} -1 & 0 \\ 0 & -1 \end{pmatrix},$$

$$l = \begin{pmatrix} \zeta^2 & 0 \\ 0 & \zeta \end{pmatrix}, \qquad a = \begin{pmatrix} 0 & -1 \\ 1 & 0 \end{pmatrix}, \qquad w = \begin{pmatrix} -\zeta & 0 \\ 0 & 1 \end{pmatrix},$$

2020 *Mathematics Subject Classification.* Primary 20-04; Secondary 20G15, 20H25, 68W30.

The author's work has been supported in part by NSF Grant DMS-1720146, and Simons Foundation Grant 852063, which are gratefully acknowledged.

subject to the relations/relators

$$tu = ut, j^2, tj = jt, uj = ju, lj = jl, aj = ja, l^3, l^{-1}tl = t^{-1}u^{-1},$$
$$l^{-1}ul = t, a^2 = j, (al)^2 = j, (ta)^3 = j, (ual)^3 = j, wj = jw, w^6, wtw^{-1} = u^{-1},$$
$$wuw^{-1} = tu, waw^{-1} = jl^2a, wl = lw.$$

An application of Tietze transformations [**6**, §5.3.3], eliminating the redundant generators $u = wt^{-1}w^{-1}$, $j = a^2$, $l = w^{-1}a^{-1}wa^{-1}$, gives us $\Gamma = \langle t, a, w\rangle$.

Using a norm-based reduction, as described in [**7**], we can obtain expressions for the generators of S as words in the generators of Γ as:

$$\begin{aligned}
m_1 &= w * (tawt)^{-}8/w,\\
m_2 &= w^{-1}aw^{-1}(t^{-1}w^{-1}t^{-1}a^{-1}twta^{-1})^3t^{-1}w^{-1}t^{-1}a^{-1}twt,\\
m_3 &= w^{-1}a^{-1}w^{-1}(a^{-1}twta^{-1}t^{-1}w^{-1}t^{-1})^4a^{-1},\\
m_i &= a^{-1},\\
m_j &= w^2a^{-1}t^{-1}w^{-1}a^{-1}t^{-1}/w\\
m_t &= (w/a)^2.
\end{aligned}$$

Attempts to determine the index if S in Γ by coset enumeration fail.

Following the approach of [**3**], we next look at congruence images. Since Γ does not satisfy the congruence subgroup property this will only ever provide a lower bound for the index of the subgroup, even if it is finite.

Let φ be the reduction modulo 2 map $\Gamma \to \mathrm{GL}_2(\mathbb{F}_4)$. We find that $|\varphi(S)| = 12$, while $|\varphi(\Gamma)| = 180$, showing that $[\Gamma : S] \geq 15$ and in particular that $\Gamma \neq S$.

Working simultaneously modulo $3, 7, 31, 97, 169, 361, 607$ (primes, found experimentally, modulo which a 3rd root of unity exists, and modulo which the image of S has index at least $p/2$ in the respective $\mathrm{GL}_2(p)$) finds a quotient of Γ of order

$$2^{43}3^{15}5^37^413^519^631{\cdot}97{\cdot}101^2607$$

in which the image of S has index $2^{27}3^{10}5^27^313{\cdot}19^231{\cdot}97{\cdot}101^2607$.

An index so large clearly puts coset enumeration outside the range of feasibility. The fact that it has been so easy to build up so large an index (indeed we could have tested further primes and would have obtained an even larger index), however indicates that the subgroup might in fact be of infinite index, and this is what we will show now:

The method we shall use is to use a normal subgroup $N \lhd \Gamma$ of finite index, that has an infinite abelian quotient, and so that $S \cap N$ has small index in S. (The former is possible only because Γ does not satisfy the congruence subgroup property.) We then calculate generators for $N \cap S$ and show that $[N : N \cap S]$ is infinite, contradicting that S could have finite index in Γ.

Concretely, let φ be the reduction on Γ modulo 7. Let $N = \ker \varphi$, then $[G : N] = |\mathrm{GL}_2(7)| = 2016$. We also calculate $|\varphi(S)| = 24$. This means that $N \cap S$ has index $|\varphi(S)| = 24$ in S.

We now construct generators for $N \cap S$. S acts, through φ on $\mathrm{GL}_2(7)$ by right multiplication, and the identity matrix has an orbit of length $|\varphi(S)| = 24$. The stabilizer of the identity matrix is $N \cap S$, and we find generators (121 of them) of $S \cap N$ as Schreier generators.

Next we use Reidemeister-Schreier rewriting [**6**, §5.3] to obtain a presentation for N. Using this presentation, a Smith normal form calculation [**6**, §9.2] will construct a homomorphism μ on N such that $\ker \mu = N'$. In this example we find that $\mu(N) \cong \mathbb{Z}^8$. While GAP represents the infinite factor group as a finitely presented group, a simplification with Tietze transformations results in a presentation on 8 generators. In this presentation the isomorphism to $\mathbb{Z}^8$ is given simply by considering exponent sums of words.

We find that $\mu(N \cap S)$ is a submodule of $\mathbb{Z}^8$ of rank 3. Thus $[N : N \cap S] = \infty$ and we thus have shown that $[G : S] = \infty$, as claimed.

Unfortunately, this approach will not work in cases when the congruence subgroup property holds. We give a justification of this for the case of $\Gamma = \mathrm{SL}_n(\mathbb{Z})$, $n \geq 3$: A subgroup of finite index in Γ (our N in the previous argument) is a congruence subgroup, and thus the pre-image of a subgroup of a congruence image $\varphi(\Gamma)$ for some congruence map φ, say the congruence is modulo m. If N has an infinite abelian quotient, we can find quotients α_p of N that are cyclic of any prime order p, in particular for primes that are larger than any primes in the order of $\varphi(\Gamma)$. But that means that for arbitrary large primes p, Γ has congruence images, whose order is a multiple of p, without involving a composition factor $\mathrm{PSL}_n(p)$, in contradiction to the structure of congruence images of $\mathrm{SL}_n(\mathbb{Z})$.

References

[1] Henrik Bäärnhielm, Derek Holt, C. R. Leedham-Green, and E. A. O'Brien, *A practical model for computation with matrix groups*. part 1, J. Symbolic Comput. **68** (2015), no. part 1, 27–60, DOI 10.1016/j.jsc.2014.08.006. MR3283836

[2] Andreas Bächle, Sugandha Maheshwary, and Leo Margolis, *Abelianization of the unit group of an integral group ring*, Pacific J. Math. **312** (2021), no. 2, 309–334, DOI 10.2140/pjm.2021.312.309. MR4305775

[3] A. Detinko, D. L. Flannery, and A. Hulpke, *Zariski density and computing in arithmetic groups*, Math. Comp. **87** (2018), no. 310, 967–986, DOI 10.1090/mcom/3236. MR3739225

[4] A.Detinko, D. L. Flannery,and A. Hulpke, *Experimenting with symplectic hypergeometric monodromy groups*, Experiment. Math. (2020), in press.

[5] The GAP Group, *GAP – Groups, Algorithms, and Programming, Version 4.11*, `http://www.gap-system.org`, 2020.

[6] Derek F. Holt, Bettina Eick, and Eamonn A. O'Brien, *Handbook of computational group theory*, Discrete Mathematics and its Applications (Boca Raton), Chapman & Hall/CRC, Boca Raton, FL, 2005, DOI 10.1201/9781420035216. MR2129747

[7] Alexander Hulpke, *Constructive membership tests in some infinite matrix groups*, ISSAC'18—Proceedings of the 2018 ACM International Symposium on Symbolic and Algebraic Computation, ACM, New York, 2018, pp. 215–222, DOI 10.1145/3208976.3208983. MR3840384

[8] Charles C. Sims, *Computation with finitely presented groups*, Encyclopedia of Mathematics and its Applications, vol. 48, Cambridge University Press, Cambridge, 1994, DOI 10.1017/CBO9780511574702. MR1267733

[9] Richard G. Swan, *Generators and relations for certain special linear groups*, Advances in Math. **6** (1971), 1–77, DOI 10.1016/0001-8708(71)90027-2. MR284516

Department of Mathematics, Colorado State University, 1874 Campus Delivery, Fort Collins, Colorado, 80523-1874

Email address: `hulpke@colostate.edu`

Contemporary Mathematics
Volume **783**, 2023
https://doi.org/10.1090/conm/783/15736

Geometric algorithms for discreteness and faithfulness

Michael Kapovich

ABSTRACT. In this paper I describe two geometric algorithms for certifying discreteness and freeness of finitely generated subgroups of $O(n,1)$, $SL(n,\mathbb{R})$ and, more generally, algorithms for discreteness and faithfulness of certain linear representations of finitely-presented groups.

1. Introduction

The goal of this paper is to describe two algorithms[1] for certifying discreteness of finitely generated subgroups of $SL(n,\mathbb{R})$ and, more generally, algorithms for discreteness and faithfulness of certain representations of finitely-presented groups. The fundamental questions that these algorithms aim to address are:

QUESTION 1.1.

1. Suppose that Γ is a hyperbolic group defined via its finite presentation. Given a homomorphism $\rho : \Gamma \to SL(n,\mathbb{R}) = G$, determine if ρ has finite kernel and discrete image.
2. Given matrices $A_1, \ldots, A_k \in G$ determine if the subgroup $\Gamma = \langle A_1, \ldots, A_k\rangle$ generated by these elements is discrete and/or free of rank k.

There is a separate issue as to what an *algorithm* even means in this setting. One approach is to work with the BSS (Blum–Schub–Smale) or *Real Ram* model of computability over the real numbers. Another approach is to assume that Γ lies in $SL(n,F)$, where F is a number field, e.g. Γ is a subgroup of an arithmetic group. We refer the reader to the papers by Jane Gilman [**G2**, **G3**] and the author, [**K1**], for discussion of the problems one is facing here. In this paper, we will ignore these foundational issues and concentrate on the geometric side of the problem. We also refer to the paper by Gilman and Maskit [**GM**] as well as other papers by Gilman, [**G1**, **G2**, **G3**] for the description and discussion of a very different geometric algorithm for discreteness of subgroups of $PSL(2,\mathbb{R})$.

In the first part of the paper, we discuss geometric algorithms dealing with the case of subgroups of $G = O^+(n,1)$ where much is known. In the second part, we discuss the general case, which is far less studied.

2020 *Mathematics Subject Classification.* Primary 22E40, 20-08; Secondary 20F67, 53C35.
Key words and phrases. Discrete groups, symmetric spaces.

[1] I am not discussing here various versions of the *ping-pong* argument, since this argument is widely known.

2. Basic hyperbolic geometry

All the material of this and the next two sections is standard; proofs can be found for instance in Ratcliffe's book [**Ra**].

We let V be an $n+1$-dimensional real vector space equipped with the Lorentzian bilinear form $\langle\cdot,\cdot\rangle$ of signature $(n,1)$. Concretely, $V=\mathbb{R}^{n+1}$ and

$$\langle \mathbf{x},\mathbf{y}\rangle = -x_0y_0 + x_1y_1 + \cdots + x_ny_n.$$

We let V^- denote the subset of V consisting of future-directed (i.e. satisfying $\langle \mathbf{x}, e_0\rangle > 0$) and negative (also known as *time-like*, i.e. satisfying $\langle \mathbf{x},\mathbf{x}\rangle < 0$) vectors in V. We will be identifying the hyperbolic n-space $\mathbb{H}^n$ with the *imaginary unit sphere* in V^-:

$$\mathbb{H} = \{\mathbf{x}\in V^- : \langle \mathbf{x},\mathbf{x}\rangle = -1\}.$$

The group $O(n,1)=O(V,\langle\cdot,\cdot\rangle)$ of linear transformations preserving the form $\langle\cdot,\cdot\rangle$ is disconnected. We let $O^+(n,1)$ denote the index two subgroup of $O(n,1)$ preserving the cone V^-. This subgroup also preserves the imaginary unit sphere $\mathbb{H}$ and equals the isometry group of the hyperbolic n-space $\mathbb{H}^n$.

By abusing the notation, we call vectors $\mathbf{x}$ satisfying $\langle \mathbf{x},\mathbf{x}\rangle = -1$, the *unit vectors*. The *tangent space* $T_{\mathbf{x}}\mathbb{H}$ to $\mathbb{H}$ at $\mathbf{x}\in\mathbb{H}$ is defined as the space of vectors orthogonal to $\mathbf{x}$:

$$T_{\mathbf{x}}\mathbb{H} = \{\mathbf{v}\in V : \langle \mathbf{x},\mathbf{v}\rangle = 0\}.$$

All nonzero vectors in this tangent space are *space-like*, i.e. satisfy

$$\langle \mathbf{v},\mathbf{v}\rangle > 0.$$

For space-like vectors we have the Lorentzian norm $|\mathbf{v}|=\sqrt{\langle \mathbf{v},\mathbf{v}\rangle}$, while for time-like vectors we also have the "norm" $|\mathbf{v}|=\sqrt{-\langle \mathbf{v},\mathbf{v}\rangle}$. The *angle* $\alpha=\angle(\mathbf{u},\mathbf{v})$ between space-like vectors is defined by the usual formula:

$$\langle \mathbf{u},\mathbf{v}\rangle = \cos(\alpha)|\mathbf{u}|\cdot|\mathbf{v}|.$$

It is also convenient to use for points in $\mathbb{H}^n$ the equivalence classes of vectors in V^-, where two vectors are equivalent if they are multiples of each other. For a vector $\mathbf{x}\in V^-$, we define its 'normalization', the unit vector

$$\bar{\mathbf{x}} = \frac{\mathbf{x}}{|\mathbf{x}|}.$$

Hyperbolic distance. For $\mathbf{x},\mathbf{y}\in V^-$ we have the following formula for the *hyperbolic distance* $d(\mathbf{x},\mathbf{y})$:

$$\cosh d(\mathbf{x},\mathbf{y}) = -\frac{\langle \mathbf{x},\mathbf{y}\rangle}{|\mathbf{x}|\cdot|\mathbf{y}|}.$$

To be more precise, this formula defines the hyperbolic distance between the normalizations $\bar{\mathbf{x}},\bar{\mathbf{y}}\in\mathbb{H}$.

The *geodesic segment* $\mathbf{xy}$ between points $\mathbf{x},\mathbf{y}\in\mathbb{H}$ is defined as

$$\mathbf{xy} = \{\bar{\mathbf{z}} : \mathbf{z} = (1-t)\mathbf{x} + t\mathbf{y}, t\in[0,1]\}.$$

Accordingly, the *midpoint* $m(\mathbf{x},\mathbf{y})$ (the point on $\mathbf{xy}$ dividing this segment in two equal parts) is given by

$$m(\mathbf{x},\mathbf{y}) = \frac{\mathbf{x}+\mathbf{y}}{|\mathbf{x}+\mathbf{y}|}. \tag{2.1}$$

We also define the (space-like) vector $\mathbf{v} = \mathbf{v}(\mathbf{x}, \mathbf{y})$ as a (not necessarily unit) vector tangent to $\mathbf{xy}$ at the point $\mathbf{x}$:

$$\mathbf{v} = \mathbf{v}(\mathbf{x}, \mathbf{y}) = \mathbf{y} + \langle \mathbf{x}, \mathbf{y} \rangle \mathbf{x}.$$

Then

$$|\mathbf{v}|^2 = \langle \mathbf{x}, \mathbf{y} \rangle^2 - 1.$$

Hyperbolic angle. The *hyperbolic angle* $\alpha = \angle_x(\mathbf{u}, \mathbf{v})$ between two nonzero tangent vectors $\mathbf{u}, \mathbf{v} \in T_{\mathbf{x}}\mathbb{H}$ at $\mathbf{x} \in \mathbb{H}$ is given, as above, by the formula

$$\langle \mathbf{u}, \mathbf{v} \rangle = \cos(\alpha)|\mathbf{u}| \cdot |\mathbf{v}|.$$

Lastly, the hyperbolic angle $\alpha = \angle \mathbf{yxz}$ of a hyperbolic triangle $\mathbf{xyz} \subset \mathbb{H}$ at the vertex $\mathbf{x}$ is defined by the formula

$$\alpha = \angle(\mathbf{v}(\mathbf{x}, \mathbf{y}), \mathbf{v}(\mathbf{x}, \mathbf{z})),$$

i.e.

$$\cos(\alpha) = \frac{\langle \mathbf{v}(\mathbf{x}, \mathbf{y}), \mathbf{v}(\mathbf{x}, \mathbf{z}) \rangle}{|\mathbf{v}(\mathbf{x}, \mathbf{y})| \, |\mathbf{v}(\mathbf{x}, \mathbf{z})|} = \frac{\langle \mathbf{y}, \mathbf{z} \rangle + \langle \mathbf{x}, \mathbf{y} \rangle \langle \mathbf{x}, \mathbf{z} \rangle}{(\langle \mathbf{x}, \mathbf{y} \rangle^2 - 1)^{1/2} (\langle \mathbf{x}, \mathbf{z} \rangle^2 - 1)^{1/2}}$$

REMARK 2.1. Computing (or, at least, estimating from below) $\sin(\alpha/2)$ will be useful for the algorithms described in sections 8, 9.

For instance, $\alpha \geqslant \pi/2$ if and only if

$$\langle \mathbf{y}, \mathbf{z} \rangle + \langle \mathbf{x}, \mathbf{y} \rangle \langle \mathbf{x}, \mathbf{z} \rangle \leqslant 0.$$

Projectivizing the negative cone V^- one obtains the *Klein model* of the hyperbolic space $\mathbb{H}^n$, which is an open ball B in the projective space $\mathbb{R}P^n = PV$: The ball B is the projection $\pi(V^-)$ of V^- to the projective space PV. The boundary sphere S^{n-1} of this ball is the projectivization of the null-cone $\{\mathbf{v} \in V : \langle \mathbf{v}, \mathbf{v} \rangle = 0\}$ of *light-like* vectors, also known as *null-vectors*. The hyperboloid $\mathbb{H} \subset V^-$ projects diffeomorphically to B, making B a model of the hyperbolic n-space.

Let $\xi = \pi(\mathbf{v})$, be a point of the boundary sphere S^{n-1} of the ball B. Without loss of generality we can assume that $\mathbf{v}$ is a nonzero vector which belongs to the closure of the cone V^-. We next define *horospheres* $\Sigma \subset \mathbb{H}$ *based* at ξ. Given $\mathbf{v}$, consider the affine hyperplane

$$\{\mathbf{x} \in V : \langle \mathbf{x}, \mathbf{v} \rangle = 1\}.$$

Replacing $\mathbf{v}$ by its positive multiple results in another hyperplane, parallel to the one defined before. Intersecting such hyperplanes with $\mathbb{H}$ yields a foliation of $\mathbb{H}$ by hypersurfaces, called *horospheres* $\Sigma \subset \mathbb{H}$ based at ξ.

3. Hyperbolic bisectors

Given two distinct points $\mathbf{p}, \mathbf{q} \in \mathbb{H}^n$, the *bisector* $Bis(\mathbf{p}, \mathbf{q})$ of the geodesic segment $\mathbf{pq}$ in $\mathbb{H}^n$ is the collection of all points which are equidistant from $\mathbf{p}$ and $\mathbf{q}$:

$$Bis(\mathbf{p}, \mathbf{q}) = \{\mathbf{x} \in \mathbb{H}^n : d(\mathbf{p}, \mathbf{x}) = d(\mathbf{q}, \mathbf{x})\}.$$

In terms of Lorentzian geometry, when $\mathbf{p}, \mathbf{q}$ are in the hyperboloid $\mathbb{H}$,

$$Bis(\mathbf{p}, \mathbf{q}) = \mathbb{H} \cap \{\mathbf{x} \in V : \langle \mathbf{p}, \mathbf{x} \rangle = \langle \mathbf{q}, \mathbf{x} \rangle\},$$

with

$$\{\mathbf{x} \in V : \langle \mathbf{p}, \mathbf{x} \rangle = \langle \mathbf{q}, \mathbf{x} \rangle\} = (\mathbf{p} - \mathbf{q})^{\perp},$$

the latter is the Lorentzian orthogonal complement to the time-like vector $\mathbf{p}-\mathbf{q}$. Let us determine when two bisectors $Bis(\mathbf{p}, \mathbf{q}_1), Bis(\mathbf{p}, \mathbf{q}_2)$ have empty intersection in $\mathbb{H}$. The empty intersection condition is equivalent to the property that all nonzero elements of the (typically) codimension 2 linear subspace

$$(\mathbf{p}-\mathbf{q}_1)^{\perp} \cap (\mathbf{p}-\mathbf{q}_2)^{\perp}$$

are null or space-like vectors. Suppose, for a moment, that one of the vectors $\mathbf{x}$ of this intersection is time-like. Then the Gram matrix of the Lorentzian bilinear form restricted to $\mathrm{span}(\mathbf{p}-\mathbf{q}_1, \mathbf{p}-\mathbf{q}_2, \mathbf{x})$, in terms of the basis $\{\mathbf{v}_1 = \mathbf{p}-\mathbf{q}_1, \mathbf{v}_2 = \mathbf{p}-\mathbf{q}_2, \mathbf{x}\}$, equals

$$\begin{bmatrix} \langle \mathbf{v}_1, \mathbf{v}_1\rangle & \langle \mathbf{v}_1, \mathbf{v}_2\rangle & 0 \\ \langle \mathbf{v}_1, \mathbf{v}_2\rangle & \langle \mathbf{v}_2, \mathbf{v}_2\rangle & 0 \\ 0 & 0 & \langle \mathbf{x}, \mathbf{x}\rangle \end{bmatrix},$$

where $\langle \mathbf{x}, \mathbf{x}\rangle < 0, \langle \mathbf{v}_i, \mathbf{v}_i\rangle > 0, i = 1, 2$. Since the signature of the restriction cannot be $(1, 2)$, the submatrix

$$\begin{bmatrix} \langle \mathbf{v}_1, \mathbf{v}_1\rangle & \langle \mathbf{v}_1, \mathbf{v}_2\rangle \\ \langle \mathbf{v}_1, \mathbf{v}_2\rangle & \langle \mathbf{v}_2, \mathbf{v}_2\rangle \end{bmatrix}$$

has to be positive semidefinite, which translates to the inequality

$$\langle \mathbf{v}_1, \mathbf{v}_1\rangle\langle \mathbf{v}_2, \mathbf{v}_2\rangle - \langle \mathbf{v}_1, \mathbf{v}_2\rangle^2 \geqslant 0.$$

The inequality is strict unless $\mathbf{v}_1 = \mathbf{v}_2$, i.e. $\mathbf{q}_1 = \mathbf{q}_2$, which we assume not to be the case. We thus arrive to:

LEMMA 3.1. *The intersection $Bis(\mathbf{p}, \mathbf{q}_1) \cap Bis(\mathbf{p}, \mathbf{q}_2)$ is empty if and only if*

$$\langle \mathbf{v}_1, \mathbf{v}_1\rangle\langle \mathbf{v}_2, \mathbf{v}_2\rangle - \langle \mathbf{v}_1, \mathbf{v}_2\rangle^2 \leqslant 0.$$

Furthermore, the intersection $(\mathbf{p}-\mathbf{q}_1)^{\perp} \cap (\mathbf{p}-\mathbf{q}_2)^{\perp}$ consists entirely of space-like vectors (and the vector 0) if and only if

$$\langle \mathbf{v}_1, \mathbf{v}_1\rangle\langle \mathbf{v}_2, \mathbf{v}_2\rangle - \langle \mathbf{v}_1, \mathbf{v}_2\rangle^2 < 0.$$

4. Isometries of the hyperbolic space

Isometries of the hyperbolic space $\mathbb{H}$ are elements of the group $O^+(n, 1)$: They are the linear transformations preserving $\mathbb{H}$. As many things in this world, isometries of the hyperbolic space fall into three groups:

- **Hyperbolic** (also called loxodromic). Every hyperbolic matrix A has two[2] distinct positive real eigenvalues $\lambda > 1, \lambda^{-1} \in (0, 1)$, of multiplicity one, and corresponding eigenvectors $\mathbf{v}_{\pm}$ which are null-vectors in V. These vectors span a 2-dimensional subspace H_A in V, whose intersection with $\mathbb{H}$ is a complete hyperbolic geodesic h_A (a hyperbola in H_A), called *the axis* of A: This geodesic is preserved by A. These isometries will be most important for us.
- **Parabolic**. These isometries A have exactly one (up to scaling) null eigenvector in V and it is fixed by A. (In other words, each parabolic isometry has exactly one fixed point in the boundary sphere S^{n-1} of the Klein model of $\mathbb{H}^n$.) All eigenvalues of A have absolute value 1.

[2]Besides these two real eigenvalues there will be other (complex) eigenvalues, but they all have absolute value 1.

- **Elliptic.** This is everything else, but one can also define these isometries A by the condition that they have fixed vectors in V^- (equivalently, in $\mathbb{H}$). All eigenvalues of A again have absolute value 1.

Taking powers does not change the type of an isometry, but taking products, in general, does. Each parabolic isometry fixes a unique point $\xi \in S^{n-1}$ and can be shown to preserve each horosphere in $\mathbb{H}^n$ based at ξ. In contrast, hyperbolic isometries of $\mathbb{H}^n$ do not preserve any horospheres. An elliptic isometry has an invariant horosphere if and only if it fixes a point in S^{n-1}. It then preserves all horospheres based at that point.

Displacement. Each hyperbolic matrix A acts on its axis h_A as a *translation*[3] by some number τ_A, i.e.

$$d(A\mathbf{x}, \mathbf{x}) = \tau_A$$

for all $\mathbf{x} \in h_A$. The number τ_A is computable in terms of the eigenvalues of A:

$$\cosh(\tau_A) = \frac{\lambda + \lambda^{-1}}{2}.$$

Checking hyperbolicity is easy: One computes the eigenvalues and checks if one of them is > 1. This works especially nicely for elements in arithmetic subgroups since there is a uniform lower bound on $\tau(A)$ for hyperbolic elements A of such groups, defined in terms of the arithmetic data of Γ. Conjecturally, for arithmetic subgroups there is a uniform positive lower bound on the displacements, depending only on the dimension:[4]

CONJECTURE 4.1. *For every $n \geqslant 2$ there exists $t(n) > 0$ such that for every arithmetic lattice $\Gamma < O^+(n,1)$, every hyperbolic element $\gamma \in \Gamma$ satisfies $\tau_\gamma > t(n)$.*

Note that this conjecture fails if we do not restrict to arithmetic lattices.

5. Connecting hyperbolic geodesics

In this section I discuss how to connect two hyperbolic geodesics h_1, h_2 in $\mathbb{H}^n$ by a geodesic segment s meeting both orthogonally at its end-points. Such a segment is the shortest segment connecting the two geodesics (unless the geodesics intersect in $\mathbb{H}^n$, which we will assume not to be the case in what follows). The segment s is known to be unique.[5] The segment s exists, unless two geodesics h_1, h_2 are asymptotic to a common point in the boundary sphere of $\mathbb{H}^n$. We will be assuming that geodesics h_i are given as the intersections of $\mathbb{H}$ with the linear 2-dimensional subspaces $W_i = \text{span}(\mathbf{u}_i, \mathbf{v}_i)$, where $\mathbf{u}_i, \mathbf{v}_i$ are (linearly independent) null-vectors in the future null-cone. Then, testing for the existence of a common asymptotic point in S^{n-1} is easy: One simply verifies if two of the four null-vectors $\{\mathbf{u}_1, \mathbf{v}_1, \mathbf{u}_2, \mathbf{v}_2\}$ are multiples of each other. Checking if $h_1 \cap h_2$ is nonempty is also easy: One computes the intersection $W_1 \cap W_2$ (which, generically, if $n \geqslant 3$, will be zero) and checks if this intersection contains a (nonzero) time-like vector.

REMARK 5.1. Of prime importance for us is the case when $W_i = H_{A_i} = \text{span}(\mathbf{u}_i, \mathbf{v}_i)$, where A_i are hyperbolic elements of $O(n,1)$ and $\mathbf{u}_i, \mathbf{v}_i$ are their light-like future-directed eigenvectors.

[3] After we identify h_A with the real line by the hyperbolic arc-length parameterization.

[4] This is a special case of the *Lehmer Problem* and *Margulis Conjecture*.

[5] This uniqueness comes from the non-existence of rectangles in spaces of negative curvature.

We first work out a condition for orthogonality of two geodesics in $\mathbb{H}^n$ meeting at a common point, as is the case for, say, h_1 and the unique hyperbolic geodesic containing s. Fix $\mathbf{y}_1 = t_1\mathbf{u}_1 + (1-t_1)\mathbf{v}_1$, $t_1 \in (0,1)$, a time-like vector in W_1. (Its normalization $\bar{\mathbf{y}}_1$ is a point in $h_1$and all points in h_1 appear this way.) Then in W_1 we find a (space-like) vector $\mathbf{y}_1^*$ orthogonal to $\mathbf{y}_1$, given by

$$\mathbf{y}_1^* = t_1\mathbf{u}_1 + (t_1 - 1)\mathbf{v}_1.$$

Suppose now that W is the plane spanned by a time-like vector $\mathbf{y}_1 = t_1\mathbf{u}_1 + (1-t_1)\mathbf{v}_1$ (as above) and another time-like vector $\mathbf{y}_2$. Then the hyperbolic geodesics h_1, h in $\mathbb{H}$ defined as $h_1 = W_1 \cap \mathbb{H}, h = W \cap \mathbb{H}$, are orthogonal if and only if the vector $\mathbf{y}_1^* \in W_1$ (Lorentzian-orthogonal to $\mathbf{y}_1$) is also Lorentzian-orthogonal to the entire plane W, equivalently, is Lorentzian-orthogonal to $\mathbf{y}_2$. The latter orthogonality condition is

$$\langle t_1\mathbf{u}_1 + (1-t_1)\mathbf{v}_1, \mathbf{y}_2\rangle = 0.$$

This gives us a way to compute the hyperbolic segment s connecting h_1, h_2 orthogonally. Namely, we search for vectors $\mathbf{y}_i = t_i\mathbf{u}_i + (1-t_i)\mathbf{v}_i$, $t_i \in (0,1)$, $i = 1, 2$, satisfying the two equations:

$$\langle \mathbf{y}_1^*, \mathbf{y}_2\rangle = \langle \mathbf{y}_2^*, \mathbf{y}_1\rangle = 0,$$

equivalently,

$$\langle t_1\mathbf{u}_1 + (t_1-1)\mathbf{v}_1, t_2\mathbf{u}_2 + (1-t_2)\mathbf{v}_2\rangle = \langle t_2\mathbf{u}_2 + (t_2-1)\mathbf{v}_2, t_1\mathbf{u}_1 + (1-t_1)\mathbf{v}_1\rangle = 0.$$

Then the common perpendicular segment to h_1, h_2 equals $s = \bar{\mathbf{y}}_1\bar{\mathbf{y}}_2 \subset \mathbb{H}$. Searching for vectors $\mathbf{y}_1, \mathbf{y}_2$ amounts to solving the above system of two quadratic equations with the unknowns t_1, t_2.

Lastly, we consider the special case $n = 2$, i.e. the vector space V is 3-dimensional, when the search problem simplifies. Then W_1, W_2 are defined as

$$W_i = \mathbf{p}_i^\perp, i = 1, 2,$$

the Lorentzian orthogonal complements to some space-like vectors $\mathbf{p}_1, \mathbf{p}_2$ in V. (In order to determine these vectors, one solves the linear systems $\langle \mathbf{p}_i, \mathbf{u}_i\rangle = \langle \mathbf{p}_i, \mathbf{v}_i\rangle = 0$, $i = 1, 2$.) The segment $s = \bar{\mathbf{y}}_1\bar{\mathbf{y}}_2 \subset \mathbb{H}$ is contained in a hyperbolic geodesic h defined as $\mathbb{H} \cap W$,

$$W = \mathbf{p}^\perp,$$

where $\mathbf{p} \in V$ is a nonzero vector satisfying $\langle \mathbf{p}, \mathbf{p}_1\rangle = 0, \langle \mathbf{p}, \mathbf{p}_2\rangle = 0$. Then the (future-directed) vectors $\mathbf{y}_i$ are basis vectors of the lines $\mathbf{p}^\perp \cap W_i$, $i = 1, 2$.

6. Quasigeodesics

While the general definition is more complicated, for the computational purposes, one can think of *quasigeodesics* in $\mathbb{H}^n$ as certain special piecewise-geodesic paths c in $\mathbb{H}^n$; their advantage over geodesics is that they are more combinatorial objects. Each quasigeodesic comes with a certain constant $\lambda(c) \geqslant 1$, the *quasiisometry constant* of c, defined by the condition that for any two points p, q on c, we have

$$d(p,q) \geqslant \lambda^{-1}|c_{p,q}| - \lambda,$$

where $|c_{p,q}|$ is the length of the subpath of c between p and q.

REMARK 6.1. The above definition of quasigeodesics is not the most general, but it is the most appropriate for computational purposes and suffices when dealing with group-homomorphisms. For the general treatment of quasigeodesics we refer the reader to [**DK**].

The constant λ measures 'how far c is from being a geodesic.' The magic of hyperbolic geometry is that every quasigeodesic, even an infinite one, is within uniformly bounded distance[6] from some geodesic. (This property is known as the *Morse Lemma*.)

The first paragraph is, in fact totally irrelevant for the computational purposes, it is mostly meant to introduce the terminology 'quasigeodesic.' The following is a practical criterion for something being a quasigeodesic, a proof could be found in [**KLi**]:

THEOREM 6.2. *Suppose that c is a piecewise-geodesic path whose angles at the vertices are $\geqslant \alpha > 0$ and whose sides are longer than L, where α and L satisfy*

$$\cosh(L/2)\sin(\alpha/2) \geqslant \nu$$

where $\nu > 1$ is some fixed constant, say, $\sqrt{2}$. Then c is a quasigeodesic. The constant $\lambda(c)$ depends only on L and ν.

REMARK 6.3. The inequality in this theorem takes a particularly simple form if $\alpha \geqslant \pi/2$ (i.e. is obtuse):

$$\cosh(L/2) \geqslant 2,$$

if we take $\nu = \sqrt{2}$ in this theorem.

The actual geometric requirement in this theorem (stated without invoking angles and lengths) is that for every two consecutive segments $\mathbf{p}_1\mathbf{p}_2, \mathbf{p}_2\mathbf{p}_3$ in the piecewise-geodesic path c, the bisectors $Bis(\mathbf{p}_1, \mathbf{p}_2), Bis(\mathbf{p}_2, \mathbf{p}_3)$ are disjoint and, moreover, are at least some fixed positive distance apart from each other. In view of Lemma 3.1, this condition translates to the language of Lorentzian geometry as:

$$\langle \mathbf{p}_2 - \mathbf{p}_1, \mathbf{p}_2 - \mathbf{p}_3\rangle^2 - \langle \mathbf{p}_2 - \mathbf{p}_1, \mathbf{p}_2 - \mathbf{p}_1\rangle\langle \mathbf{p}_2 - \mathbf{p}_3, \mathbf{p}_2 - \mathbf{p}_3\rangle \geqslant \epsilon, \tag{6.1}$$

where $\epsilon > 0$ is a fixed positive number.

Note that the inequalities that one needs to check are purely local, we just need to examine consecutive pairs of segments to verify them. This, of course, is still not feasible if there are many (or infinitely many) segments, but for quasigeodesics coming from group theory, there are only finitely many (one can even estimate how large is 'many') options for side-lengths and angles, so the verification becomes a finite problem. This is the 'local-to-global' principle in hyperbolic geometry. This principle has an analogue for higher rank Lie groups such as $SL(n, \mathbb{R})$ (and symmetric spaces they act on) but is much harder to state (and to prove). We will discuss this in section 12.

EXAMPLE 6.2 (A non-example). In the upper half-plane model $\{z \in \mathbb{C} : Im(z) > 0\}$ of the hyperbolic plane take the sequence of points

$$z_k = k + \sqrt{-1}, k \in \mathbb{Z}.$$

[6]this distance is $\leqslant a\lambda^2$, where a is some universal constant

Connect each consecutive pair z_k, z_{k+1} by a hyperbolic geodesic segment. The resulting path c has geodesic pieces of the constant length L and constant angle α between the consecutive pieces. These two numbers satisfy the equality

$$\cosh(L/2)\sin(\alpha/2) = 1,$$

but c is within infinite (hyperbolic) distance from any hyperbolic geodesic (it is uniformly close to the Euclidean horizontal line $Im(z) = 1$, a horosphere, but that does not count).

The *KLP algorithm* described later on, uses the following important property of quasigeodesic paths:

Suppose that $c = \mathbf{x}_0\mathbf{x}_1 \star \mathbf{x}_1\mathbf{x}_2 \star \mathbf{x}_3\mathbf{x}_4 \star \dots$ is a (finite or infinite) piecewise-geodesic path in $\mathbb{H}^n$, which is a concatenation of the geodesic segments $\mathbf{x}_i\mathbf{x}_{i+1}$. For a natural number N define the piecewise-geodesic path c_N as the concatenation

$$\mathbf{x}_0\mathbf{x}_N \star \mathbf{x}_{2N}\mathbf{x}_{2N}x_{3N} \star \dots$$

Then c is λ-quasigeodesic if and only if c_N is $\lambda'(\lambda, N)$-quasigeodesic, for some universal function $\lambda' : \mathbb{R}_+ \times \mathbb{N} \to \mathbb{R}_+$.

7. Group homomorphisms

Suppose that Γ is a finitely generated group with a finite generating set $S = \{s_1, \dots, s_k\}$. For concreteness, one can (and, at first, we will) assume that Γ is a free group on S. (But the discussion below will apply to other groups, things just become more complicated.)

A homomorphism

$$\rho : \Gamma \to O(n,1) = O(V, \langle \cdot, \cdot \rangle)$$

is simply a map sending the generators s_i to some matrices $A_i \in O^+(n,1)$ which satisfy the relators of Γ. One can also think of relators of Γ as 'hidden' and all what we have is a set of matrices $A_1, \dots, A_k \in O^+(n,1)$. Our task is to 'discover' the hidden relators (or to prove that there is none). The 'freeness problem' is to find a semi-algorithm ensuring that $\rho(\Gamma)$ (the subgroup of $O^+(n,1)$ generated by $A_1, \dots, A_k$) is free of rank k on the generators $A_1, \dots, A_k$.

Before doing this, we need some terminology. Given a homomorphism ρ (a choice of matrices A_i) and a vector $\mathbf{x} \in \mathbb{H}$ (which is to be chosen wisely to make computations more efficient), one defines the 'orbit map'

$$o_{\mathbf{x}} : \Gamma \to \mathbb{H},$$

sending $\gamma \in \Gamma$ to the vector $\gamma\mathbf{x}$. (Here and in what follows, I will frequently abbreviate $\rho(\gamma)\mathbf{x}$ as $\gamma\mathbf{x}$.) However, what we have is more than just a map of Γ, we also get a map $f = f_{\mathbf{x}}$ of the Cayley graph T of Γ into $\mathbb{H}$ (in the case of a free group of rank k, this graph is the $2k$-valent tree): Send each edge $e = [w_1, w_2]$ of T to the segment

$$(w_1\mathbf{x})(w_2\mathbf{x}) \subset \mathbb{H}.$$

Most importantly, each geodesic path in the tree T (an edge-path without back-tracking) is sent to a piecewise-geodesic path in $\mathbb{H}$. (For the computational purposes, one does not need to 'compute' the map f, it is used only to give a geometric explanation of what is happening.)

From the computational viewpoint, such paths in $\mathbb{H}$ are given by their vertex-sequences

$$\mathbf{x} = \mathbf{x}_0, \mathbf{x}_1, \ldots, \mathbf{x}_N,$$

defined for each reduced word $w = w(A_1^{\pm 1}, \ldots, A_k^{\pm 1})$ (of the length N) and applying inductively (in order of their appearance in w) the matrices

$$A_1^{\pm 1}, \ldots, A_k^{\pm 1}$$

to the point $\mathbf{x}$.

DEFINITION 7.1 (Undistorted homomorphisms and subgroups). A homomorphism ρ is called a *quasiisometric embedding* or, simply, *undistorted*, if the map f_x sends geodesic paths p in T to λ-quasigeodesic paths in $\mathbb{H}$ for some fixed λ independent of p. The image of an undistorted homomorphism is called an *undistorted subgroup* of $O^+(n,1)$.

One of the many (not so) magic properties of undistorted homomorphisms is that they are **faithful**, i.e. the subgroup generated by $A_1, \ldots A_k$ is free on this generating set. Moreover, this subgroup is necessarily discrete and contains only hyperbolic matrices (besides the identity). For instance, to see faithfulness, note that if ρ is not faithful then its kernel contains elements γ arbitrarily far from the neutral element of Γ. (Here is one of the few places where we use the assumption that Γ is free. General hyperbolic groups can contain nontrivial finite normal subgroups. But there is always the largest such subgroup.) But, since ρ is undistorted, we have the inequality

$$0 = d(\mathbf{x}, \rho(\gamma)\mathbf{x}) \geqslant \lambda^{-1}|\gamma| - \lambda,$$

where $|\gamma|$ is the distance from γ to the neutral element of Γ. This shows that $|\gamma| \leqslant \lambda^2$, which is a contradiction. A similar argument establishes discreteness of $\rho(\Gamma)$.

Note that the maps $o_{\mathbf{x}}$ and $f_{\mathbf{x}}$ depend on $\mathbf{x}$, but the undistortion property does not. However, an unwise choice of $\mathbf{x}$ will make the quasiisometry constant λ larger and computations longer.

There are *distorted* injective homomorphisms (with discrete images) $\rho : \Gamma \to O^+(n,1)$, even when one restricts to homomorphisms whose targets are arithmetic subgroups of $O^+(n,1)$ such as the subgroup of integer matrices $O(n,1;\mathbb{Z})$. One of the most famous examples of such homeomorphisms comes from the embedding ρ of the figure 8 knot group into $O^+(3,1)$. Algebraically, this group is a semidirect product of the free group F_2 and the infinite cyclic group. Then $\rho : F_2 \to O^+(3,1)$ is exponentially distorted. It is now known (due to work of Ian Agol) that every lattice in $O^+(3,1)$ contains a finitely generated subgroup which is either free or isomorphic to the fundamental group of a hyperbolic surface, and which is exponentially distorted in $O^+(3,1)$.

The notion of undistorted subgroups is closely related to *geometric finiteness* of subgroups, see [**Bo1**, **Bo2**]. In particular, every undistorted subgroup of $O^+(n,1)$ has finitely-sided Dirichlet fundamental polyhedra in $\mathbb{H}^n$. (We will discuss Dirichlet domains in more detail in Section 11.) The notion of *geometric finiteness* is a bit more general. For instance, an infinite cyclic subgroup of $O^+(n,1)$ generated by a parabolic element is geometrically finite but exponentially distorted. Geometric finiteness, in turn, is closely related to the property that Dirichlet fundamental polyhedra in $\mathbb{H}^n$ are finitely-sided: The two notions are equivalent if $n \leqslant 3$, but not for $n \geqslant 4$. (See Examples 5 and 6 in Section 12.4 of Ratcliffe's book [**Ra**].) However, for subgroups of lattices in $O^+(n,1)$, geometric finiteness is equivalent to the property that one (equivalently, every) Dirichlet fundamental polyhedron is finitely-sided.

For some reason, not quite clear in general, if one considers finitely generated discrete subgroups of $O^+(n,1)$, geometric finiteness appears to be a *generic property*. If $n = 2$, then every finitely generated discrete subgroup is geometrically finite. However for $n \geqslant 3$, there are finitely generated discrete geometrically infinite subgroups.

8. Testing for undistortion. Part I

If one looks closely at the maps f constructed in the previous section, one observes that f produces piecewise-geodesic paths in $\mathbb{H}$ (images of geodesics in T) that satisfy two 'finiteness' properties:

- The number of possible edge-lengths in these paths is at most k (one for each generator); they are given by the distances
$$L_1 = d(\mathbf{x}, A_1\mathbf{x}), \ldots, L_k = d(\mathbf{x}, A_k\mathbf{x}).$$
Recall that (see Section 2),
$$\cosh d(\mathbf{x}, A_i\mathbf{x}) = -\langle \mathbf{x}, A_i\mathbf{x}\rangle.$$
- The number of angles between the consecutive segments in such paths is at most $k(2k-1)$: One angle $\alpha_{i,\pm j}$ for each pair of generators $A_i, A_j^{\pm 1}$, where, of course, we do not allow pairs of the form (A_i, A_i) (which would correspond to backtracking in the tree T or, equivalently, nonreduced words w). Here
$$\alpha_{i,\pm j} = \angle(A_i\mathbf{x})\mathbf{x}(A_j^{\pm 1}\mathbf{x}),$$
that is (see Section 2),
$$\cos(\alpha_{i,\pm j}) = \frac{\langle A_i\mathbf{x}, A_j\mathbf{x}\rangle + \langle \mathbf{x}, A_i\mathbf{x}\rangle\langle \mathbf{x}, A_j^{\pm 1}\mathbf{x}\rangle}{(\langle \mathbf{x}, A_i\mathbf{x}\rangle^2 - 1)^{1/2}(\langle \mathbf{x}, A_j^{\pm 1}\mathbf{x}\rangle^2 - 1)^{1/2}}.$$

Now comes our first (and rather dumb) algorithm for testing the undistortion property. Even though it is dumb, it works quite well 'generically' and this is what's behind, say, the Fuchs–Rivin's proof of genericity of free subgroups in arithmetic groups, see [**FR**, Lemma 2.5].

The Dirty Harry Algorithm. [7] Check the inequality
$$\cosh(L/2)\sin(\alpha/2) > 1.$$
where $L = \min(L_1, \ldots, L_k)$, and
$$\alpha = \min\{\alpha_{i,\pm j} : 1 \leqslant i, j \leqslant k, (i, \pm j) \neq (i, -i)\}.$$
If this inequality holds, then indeed, f sends geodesic paths in the tree T to quasi-geodesic paths in $\mathbb{H}$ and, hence, f is undistorted and, hence, injective with discrete image.

The geometric (rather than *coarse-geometric*) meaning of the inequality in the algorithm is that it ensures that the geodesic bisectors of the segments $(\mathbf{x})(A_i^{\pm 1}\mathbf{x})$

[7] "Do you feel lucky today?"

are pairwise disjoint in $\mathbb{H}^n$ (or even in the compactified hyperbolic space). If this happens then these bisectors will bound the Dirichlet fundamental domain of $\rho(\Gamma)$ in $\mathbb{H}^n$ centered at $\mathbf{x}$. We discuss Dirichlet fundamental domains and the corresponding *Poincaré* algorithm in detail in Section 11. For now, we simply record the fact that if the Dirty Harry Algorithm succeeds, then so does the KLP-algorithm, in its $O(n,1)$-version, and the *Poincaré* algorithm. Moreover, both of the latter algorithms terminate on their first step, dealing with group elements of word-length 1.

As we discussed in Section 6, instead of computing hyperbolic distances and angles, it is easier to test disjointness of bisectors. In this, more computationally-friendly, form, the Dirty Harry Algorithm works as follows. For each generator $A_i^{\pm 1}$, compute the vector

$$\mathbf{u}_{\pm i} = \mathbf{x} - A_i^{\pm 1}\mathbf{x}.$$

Then for each pair of different vectors $\mathbf{v}, \mathbf{w} \in \{\mathbf{u}_{\pm i} : i = 1, \ldots, k\}$ compute the difference

$$D_{\mathbf{v},\mathbf{w}} := \langle \mathbf{v}, \mathbf{w} \rangle^2 - \langle \mathbf{v}, \mathbf{v} \rangle \langle \mathbf{w}, \mathbf{w} \rangle.$$

If all differences satisfy $D_{\mathbf{v},\mathbf{w}} > 0$, the algorithm succeeds and the representation ρ is discrete, faithful and even undistorted. If all the differences satisfy $D_{\mathbf{v},\mathbf{w}} \geqslant 0$, the algorithm succeeds and the representation ρ is discrete and faithful. Otherwise, i.e. if some difference $D_{\mathbf{v},\mathbf{w}}$ is negative, the algorithm fails.

Even if ρ is undistorted, the Dirty Harry Algorithm might not work. It has a better chance of success provided one makes a 'wise' choice of the point $\mathbf{x}$.

8.1. Choosing x wisely (the 2-generator case). I first consider the case of 2-generator groups. Recall that for nondistortion to occur, all nontrivial elements of $\rho(\Gamma)$ have to be hyperbolic. Therefore, one should first check for hyperbolicity of the generators A_1, A_2. If one of them is nonhyperbolic (compute the eigenvalues), one stops and proceeds to try some other matrices. Suppose that A_1, A_2 are hyperbolic.

Let $h_1 = H_{A_1} \cap \mathbb{H}, h_2 = H_{A_2} \cap \mathbb{H}$ be the axes of A_1, A_2 (see section 4). Compute the geodesic segment $s_{h_1,h_2} = \mathbf{x}_1\mathbf{x}_2$ connecting h_1 and h_2 (see section 5). Then the 'wise' choice of $\mathbf{x}$ is the midpoint $m(\mathbf{x}_1\mathbf{x}_2)$ of the segment $\mathbf{x}_1\mathbf{x}_2$ (see (2.1)).

REMARK 8.1. While there are some heuristic reasons why one should be choosing the midpoint as $\mathbf{x}$, there is no solid mathematical justification for this.

The idea of taking a midpoint extends to the case of a larger number of generators, but is computationally a bit more demanding:

Choosing x wisely (the general case). First, as above, for each pair of distinct indices i, j, compute the midpoint m_{ij} of the segment

$$s_{ij} = s_{h_i,h_j}$$

connecting the axes h_i, h_j of A_i, A_j. Then compute

$$\mathbf{b} = \sum_{i,j} m_{ij}$$

where the sum is taken over all pairs distinct elements of the set of midpoints of segments s_{ij},

$$M = \{m_{ij}, 1 \leqslant i < j \leqslant k\}.$$

Lastly, take as the point $\mathbf{x} \in \mathbb{H}$ the normalization of $\mathbf{b}$:

$$\mathbf{x} = \bar{\mathbf{b}} = \frac{\mathbf{b}}{|\mathbf{b}|}.$$

9. Testing for undistortion. Part II

I will now describe a simplified form of the KLP (Kapovich–Leeb–Porti) algorithm, written originally for subgroups of general semisimple real Lie groups, testing for the *Anosov property* of representations of hyperbolic groups; this is an adaptation and simplification in the case of $O^+(n,1)$). I will first do it for free groups and then in general.

This algorithm does not require luck: It terminates if and only if the homomorphism ρ is undistorted, therefore establishing semidecidability of the 'testing for undistortion' problem.

Instead of analyzing just the generators A_i (and their inverses), for $N \geqslant 1$ the $O^+(n,1)$-version of the KLP algorithm explores *radius N balls in the Cayley graph T centered at the neautral element* 1. If the algorithm provides the desired output for some N, it terminates, otherwise, it runs forever. Suppose that Γ is a free group with free generating set $s_1, \ldots, s_k$. Then the Cayley graph of Γ with respect to this generating set is a simplicial tree T. As before, we are given a homomorphism $\rho : \Gamma \to O^+(n,1)$, $\rho(s_i) = A_i$, $i = 1, \ldots, k$.

First, some terminology. The *radius R* (where R is a natural number) ball $B(R)$ centered at 1 in the vertex-set of the tree T is just the set of reduced words in $s_i^{\pm 1}$ of length at most R. We will be working with *(geodesic) N-triples* in such balls: These are triples of reduced words $w_1, 1, w_2$ which lie on a common geodesic segment (connecting w_1 to w_2) in the ball $B(N)$ and satisfy

$$|w_1| = |w_2| = N.$$

Here $|w|$ is the word-length of a reduced word w.

Concretely, being a geodesic N-triple means two things:

(1) $|w_1| = N, |w_2| = N$.
(2) The prefix of the word w_1 is different from the prefix of a word w_2, where the prefix of a word is the first letter (in the alphabet $s_i^{\pm 1}$, $i = 1, \ldots, k$) of the word.

Fix a point $\mathbf{x} \in \mathbb{H}$. (As before, it is best if this point is chosen wisely, but, unlike in the Dirty Harry case, the outcome of the KLP algorithm does not depend on the choice.)

Given a (geodesic) triple $\tau = (w_1, 1, w_2)$, compute the difference

$$D_\tau = \langle \mathbf{x} - w_1\mathbf{x}, \mathbf{x} - w_2\mathbf{x} \rangle^2 - \langle \mathbf{x} - w_1\mathbf{x}, \mathbf{x} - w_1\mathbf{x} \rangle \langle \mathbf{x} - w_2\mathbf{x}, \mathbf{x} - w_2\mathbf{x} \rangle.$$

DEFINITION 9.1. We say that a triple $\tau = (w_1, 1, w_2)$ satisfies the **qi condition** if $D_\tau > 0$.

Note that the triple $(w_1, 1, w_2)$ satisfies the qi condition if and only if the triple $(w_2, 1, w_1)$ does.

Now, we are ready for the actual algorithm (adapted from [**KLP1**]).

The rank one KLP algorithm. For each natural number N, consider all[8] (geodesic) N-triples $(w_1, 1, w_2)$ where w_1, w_2 are reduced words in the generators A_i, A_j^{-1},

$$|w_1| = N, |w_2| = N,$$

and the prefix of w_1 is different from the prefix of w_2.

For every such N-triple, check if it satisfies the **qi condition** as defined above. If all such N-triples pass the qi test, the algorithm stops: This means that the subgroup $\rho(\Gamma) < O^+(n,1)$ generated by $A_1 = \rho(s_1), \ldots, A_k = \rho(s_k)$ is undistorted and is free of rank k.

If one of the N-triples $(w_1, 1, w_2)$ fails the test, then stop the analysis of N-triples, increase N by 1 and repeat. As a bonus, once the algorithm stops (if it does!) we can also estimate from above the quasiisometry constant of the orbit map $o_{\mathbf{x}} : \Gamma \to \gamma\mathbf{x} \subset X$.

REMARK 9.1. Step 1 of the rank one KLP algorithm (i.e. $N = 1$) is nothing but the Dirty Harry Algorithm.

THEOREM 9.2. *The KLP algorithm terminates if and only if the subgroup* $\rho(\Gamma) < O^+(n,1)$ *generated by* $A_1, \ldots, A_k$ *is undistorted and is free of rank* k.

PROOF. The proof of this theorem is a special case of the one given in [**KLP1**, section 7]. Namely, suppose that the algorithm terminates. Then the orbit map $o_{\mathbf{x}} : \Gamma \to \mathbb{H}^n$ satisfies the following property: For each geodesic path p in T starting at 1, the restriction of $o_{\mathbf{x}}$ to p_N is λ-quasigeodesic for some λ independent of p, hence, the restriction to p is a λ'-quasigeodesic for some uniform constant λ, see the last paragraph of section 6. In order to conclude that $o_{\mathbf{x}}$ is a quasiisometry, we have to consider general paths p in T, not necessarily starting at the neutral element 1. However, the map $o_{\mathbf{x}}$ is ρ-equivariant:

$$o_{\mathbf{x}}(\gamma z) = \rho(\gamma) o_{\mathbf{x}}(z), z \in T.$$

Since Γ acts transitively on the vertex-set of T and the post-composition with isometries in $\rho(\Gamma)$ does not change the quasiisometry properties of a path, it follows that $o_{\mathbf{x}}$ is a quasiisometry.

For the opposite implication, we refer the reader to [**KLP1**]. The key is the following property: Since ρ is undistorted, for each geodesic path p in T starting at the neutral element, the vertices in $\mathbb{H}^n$ of the image c_N of p_N are uniformly close to the hyperbolic geodesic connecting $\mathbf{x}$ to the terminal point of c_N. At the same time, the distance between the consecutive points of c_N diverges to ∞ (this again uses nondistortion). Therefore, for each triple of consecutive points of c_N, say,

$$\mathbf{x}, \mathbf{x}_1 = c(N), \mathbf{x}_2 = c(2N),$$

the distances $d(\mathbf{x}, \mathbf{x}_1), d(\mathbf{x}_1, \mathbf{x}_2)$ grow arbitrarily large (as $N \to \infty$), while the angle $\angle \mathbf{x}\mathbf{x}_1\mathbf{x}_2$ is uniformly bounded away from zero. From this, one sees that the triple

$$\mathbf{x}, \mathbf{x}_1 = c(N), \mathbf{x}_2 = c(2N)$$

satisfies the qi condition if N is sufficiently large. □

One can use the 'wise choice' of $\mathbf{x}$ as described in section 8. I will describe alternatives in the next section.

---

[8]Actually from two triples $(w_1, 1, w_2)$, $(w_2, 1, w_1)$ it suffices to check just one.

One of important modifications in $\mathbb{H}^n$ of the KLP test is that it suffices to work with triples rather than with *quadruples* as it is done in [**KLP1**]. In a sense, the quadruple test from [**KLP1**] is more efficient, the drawback, however, is that one has to explore significantly larger balls in the Cayley graph. Here is the description of the quadruple test from [**KLP1**], again adapted to the case of the hyperbolic space.

Instead of triples, one works with *quadruples* of reduced words $w_0 = 1, w_1, w_2, w_3$ which lie on a common geodesic segment, satisfying $d(w_i, w_{i+1}) = N$, $i = 0, 1, 2$.

Given a (geodesic) quadruple $(1, w_1, w_2, w_3)$, compute the quadruple of vectors

$$\mathbf{x}_0 = \mathbf{x}, \mathbf{x}_1 = w_1(\mathbf{x}), \mathbf{x}_2 = w_2(\mathbf{x}), \mathbf{x}_3 = w_3(\mathbf{x})$$

in $\mathbb{H}$. For each segment $\mathbf{x}_0\mathbf{x}_1, \mathbf{x}_1\mathbf{x}_2, \mathbf{x}_2\mathbf{x}_3$ compute its midpoint

$$m_1 = m(\mathbf{x}_0, \mathbf{x}_1), m_2 = m(\mathbf{x}_1, \mathbf{x}_2), m_3 = m(\mathbf{x}_2, \mathbf{x}_3).$$

DEFINITION 9.2. Fix $\epsilon > 0$. We say that a quadruple $(1, w_1, w_2, w_3)$ satisfies the ϵ-**midpoint condition** if the triple of midpoints $(\mathbf{p}_1, \mathbf{p}_2, \mathbf{p}_3) = (m_1, m_2, m_3)$ satisfies the inequality (6.1) from Section 6.

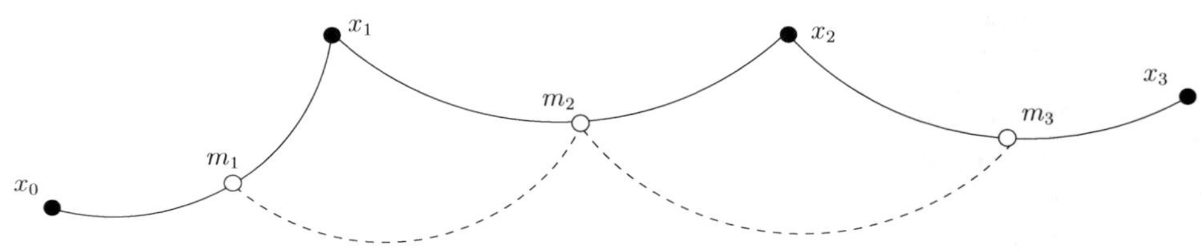

FIGURE 1. Midpoints

Then the KLP algorithm amounts to checking the midpoint condition for all (geodesic) N-quadruples $(1, w_1, w_2, w_3)$. Because one uses midpoints, in the undistorted case, the angles α are not just bounded away from zero as $N \to \infty$, but actually converge to π. (This convergence to π is critical in the higher rank case.) In particular, the products $\cosh(L/2)\sin(\alpha/2)$ diverge to infinity faster than in the qi test for triples.

10. Testing for nondistortion. Part III

Below is a version of the KLP algorithm for non-free subgroups. The algorithm is testing for the following:

Let $\Gamma = \langle a_1, \ldots, a_k | r_1, \ldots, r_s \rangle$ be a *word-hyperbolic group* given by its finite presentation. Let $\rho : \Gamma \to O^+(n, 1)$ be a homomorphism. The KLP algorithm determines if ρ is undistorted. If ρ is undistorted, it might fail to be injective, but, in the worst case, its kernel is finite. 'Most' examples of infinite hyperbolic groups have no nontrivial finite normal subgroups, thus, the nondistortion property *effectively* implies injectivity.

The only difference with section 9 is that the Cayley graph of Γ is no longer a tree and it is harder to test if a triple $(w_1, 1, w_2)$ lies on a geodesic. The right condition is

$$|w_1|_\Gamma = |w_2|_\Gamma = N, |w_1^{-1} w_2|_\Gamma = 2N$$

where $|w|_\Gamma$ is the length of the shortest word representing the same element of Γ as w. However, for hyperbolic groups (with a fixed presentation) there are practical algorithms for computing $|w|_\Gamma$ which can be used, see e.g. [**EH**]. Other than that, the KLP algorithm is the same as before.

Note that general word-hyperbolic groups, such as the triangle group $\Delta(p, q, r)$, could have generators a_i of finite order. Hence, **one cannot use** the procedure from section 8 in order to make a 'wise' choice of the vector $\mathbf{x}$. Here are the best alternatives I know:

Given matrices $A_1, \ldots, A_k$ in $O^+(n, 1)$, define the function

$$D(\mathbf{y}) := \max(d(A_1^{\pm 1}\mathbf{y}, \mathbf{y}), \ldots, d(A_k^{\pm 1}\mathbf{y}, \mathbf{y}))$$

on $\mathbb{H}$. This function is *convex* (actually, strictly convex in most examples). Then choose $\mathbf{x}$ to be the point of minimum of $D(\mathbf{y})$. In fact, it suffices just to be 'not too far' from the minimum, in any reasonable sense. From the linear algebra viewpoint, this is an unpleasant min-max problem since $D(\mathbf{y})$ is very nonlinear. Here is a practical replacement of the above min-max problem:

The min-max problem. Define the function

$$M(\mathbf{y}) := \max(-\langle A_1^{\pm 1}\mathbf{y}, \mathbf{y}\rangle, \ldots, -\langle A_k^{\pm 1}\mathbf{y}, \mathbf{y}\rangle)$$

on the open convex cone V^-. This function is piecewise-quadratic. Now, minimize this function over the hyperboloid $\mathbb{H} \subset V^-$. Choose $\mathbf{x}$ to be its minimum (even approximate one in any reasonable sense).

More alternatives. Instead of $M(\mathbf{y})$, you can take your favorite norm $\|\cdot\|$ of the k-tuple

$$(-\langle A_1^{\pm 1}\mathbf{y}, \mathbf{y}\rangle, \ldots, -\langle A_k^{\pm 1}\mathbf{y}, \mathbf{y}\rangle),$$

or even take the 'energy' (which is a quadratic function in $\mathbf{y}$)

$$E(\mathbf{y}) := -\sum_{i=1}^{k}\langle A_i\mathbf{y}, \mathbf{y}\rangle - \sum_{i=1}^{k}\langle A_i^{-1}\mathbf{y}, \mathbf{y}\rangle.$$

Now, minimize the norm, or $E(\mathbf{y})$, over the hyperboloid $\mathbb{H}$.

Other rank one Lie groups. What is described above works just as well when instead of the group $O^+(n, 1)$ of isometries of the hyperbolic n-space one considers isometry groups of other negatively curved symmetric spaces X, e.g. the group $PU(n, 1)$ of biholomorphic isometries of the complex-hyperbolic n-space. The inequality in Theorem 6.2 still implies the quasigeodesic condition provided one normalizes the Riemannian metric of X to have the upper curvature bound -1.

11. Selberg's higher rank generalization of Dirichlet domain and the Poincaré Algorithm

We begin by defining Dirichlet domains for discrete group actions on general metric spaces. Let (X, d) be a metric space and Γ a discrete isometry group of X, where discreteness is understood in the sense that for one (equivalently, every) $x \in X$ and every sequence of distinct elements $\gamma_i \in \Gamma$, we have

$$\lim_{i\to\infty} d(x, \gamma_i x) = \infty.$$

Suppose that $o \in X$ is a point which is fixed only by the neutral element of Γ. (Without this assumption, we will not obtain a fundamental domain by applying the Dirichlet construction.) Then define

$$D(\Gamma, o) = D(o) := \{x \in X : d(o, x) \leqslant d(\gamma o, x), \quad \forall \gamma \in \Gamma\}.$$

In order to understand where this definition comes from, consider the orbit, $\Gamma o \subset X$. The discreteness condition on Γ implies that this orbit is a discrete closed subset of X (moreover, each metric ball in X contains only finitely many orbit points). Thus, one defines the *Voronoi tiling* of X corresponding to this orbit:

$$D(\gamma o) = \{x \in X : d(\gamma o, x) \leqslant d(o', x) \quad \forall o' \in \Gamma o\},$$

is the tile labeled by the point γo. It is clear that every point of X belongs to one of the tiles and that Γ permutes the tiles simply-transitively. The *open tile* $\mathring{D}(\gamma o)$ is defined by

$$\mathring{D}(\gamma o) = \{x \in X : d(\gamma o, x) < d(o', x) \quad \forall o' \in \Gamma o \backslash \{\gamma o\}\}.$$

In view of continuity of the distance function $d : X^2 \to \mathbb{R}$, each $\mathring{D}(\gamma o)$ is an open subset of X. In general, however, the closure of an open tile need not be the corresponding closed tile $D(\gamma o)$. For instance, if the metric d is discrete, this will not be the case as the open tile is the singleton $\{\gamma o\}$. Moreover, the triangle inequalities imply that $\mathring{D}(\gamma o) \cap \mathring{D}(\gamma' o) \neq \varnothing$ if and only if $\gamma o = \gamma' o$. Suppose now that, additionally, (X, d) is a *geodesic space*, i.e. for any two points $x, y \in X$, $d(x, y)$ is the length of the shortest (geodesic) path in X connecting x and y. Then it is not hard to see that $\mathring{D}(\gamma o)$ is dense in $D(\gamma o)$, see [**K2**]. Thus, under this extra assumption, if $\gamma o \neq \gamma' o$, then

$$D(\gamma o) \cap D(\gamma' o) \subset \partial D(\gamma o) \cap \partial D(\gamma' o).$$

Discreteness of Γ implies that every bounded subset of X has nonempty intersection only with finitely many tiles. Thus, $D(\Gamma, o)$ serves as a fundamental domain for the action of Γ on X. We refer to [**K2**] for details.

For general metric spaces and even general Riemannian manifolds, very little can be said about geometry of Dirichlet domains and even of the *bisectors*

$$Bis(o, \gamma o) = \{x : d(o, x) = d(\gamma o, x)\}$$

bounding these domains. There is one case, however, when bisectors and, accordingly, Dirichlet domains, have particularly nice structure, namely, when (X, d) is the hyperbolic n-space $\mathbb{H}$. Then each bisector

$$Bis(\mathbf{p}, \mathbf{q}) = \{\mathbf{x} \in \mathbb{H} : \langle \mathbf{p}, \mathbf{x} \rangle = \langle \mathbf{q}, \mathbf{x} \rangle\}, \mathbf{p} \neq \mathbf{q},$$

is the intersections of the hyperboloid $\mathbb{H}$ with the linear hyperplane

$$\{\mathbf{x} \in V : \langle \mathbf{p} - \mathbf{q}, \mathbf{x} \rangle = 0\}.$$

Accordingly, $D(\Gamma, \mathbf{p})$ is the intersection of a (possibly infinitely-sided) convex polyhedral cone with $\mathbb{H}$. This polyhedral structure of $D(\Gamma, \mathbf{p})$ makes it amenable to algorithmic computations. The corresponding *Poincaré Algorithm* was described first, to my knowledge, by Riley in [**R1**] in the case $n = 3$ (who even wrote a code, in Fortran), and, in greater detail (but without actual computer implementation), by Epstein and Petronio in [**EP**].

We now specialize to the case of discrete subgroups of $G = SL(n, \mathbb{R})$, focusing on computational aspects. The group G acts naturally on the vector space V of symmetric $n \times n$ matrices, $M \mapsto g^T M g$, for $M \in V$. This action corresponds to the change of variables in the quadratic form defined by M. The group G also preserves the open cone $P \subset V$ of (strictly) positive-definite matrices and the hypersurface $X \subset P$ consisting of matrices of unit determinant. Moreover, G acts transitively on X with the stabilizer of the identity matrix equal to $SO(n)$. The hypersurface X does have a G-invariant Riemannian metric which, on the tangent space at the identity matrix I, equals

$$\langle A, B \rangle = tr(AB).$$

Given this metric, one defines the associated Riemannian distance function d. For instance,

$$d(I, A) = \left(\sum_i \log^2(\lambda_i) \right)^{1/2},$$

where λ_i's are the eigenvalues of $A \in X$.

Then, given a discrete subgroup $\Gamma < G$ which has trivial intersection with $SO(n)$, one defines the *Dirichlet domain* as above by

$$D(\Gamma, I) = \{C \in X : d(I, C) \leqslant d(\gamma^T \gamma, C), \forall \gamma \in \Gamma\}.$$

The trouble is that such domains are bounded by pieces of *geodesic bisectors* for the metric d, which are hard to compute (unlike in the case of the Lorentzian model of the hyperbolic space, where bisectors are linear).

Below, we describe a 2-point invariant $s(A, B)$ due to Selberg, [**Se**], which, while not a metric, can be used in lieu of one to define Dirichlet domains in P (and in X). This use is also due to Selberg but appears to be relatively unknown. The advantage of $s(A, B)$ is that it is easy to compute and the corresponding bisectors are linear.

Assuming that $A \in V$ is an invertible matrix and $B \in V$ is an arbitrary matrix, we define $s(A, B) := tr(A^{-1}B)$. Assuming further that $A, B \in P$ (at this point, we do not yet impose the condition $A, B \in X$), we set

$$\sigma(A, B) := \log\left(\frac{1}{n} s(A, B)\right) = \log\left(\frac{1}{n} tr(A^{-1}B)\right).$$

Then s, and, hence, σ, is G-invariant because trace is conjugacy-invariant:

$$s(g^T A g, g^T B g) = tr(g^{-1}A^{-1}(g^T)^{-1}g^T B g) = tr(g^{-1}A^{-1}Bg) = tr(A^{-1}B) = s(A, B).$$

The normalization (in the definition of σ) is chosen so that for $A, B \in X$ (i.e. $\det(A) = \det(B) = 1$), $\sigma(A, B) \geqslant 0$ with equality if and only if $A = B$, making σ a *premetric*: Indeed, without loss of generality, we may assume that both matrices A, B are diagonal,

$$A = Diag(a_1, \ldots, a_n), \quad B = Diag(b_1, \ldots, b_n).$$

Then, by the AM–GM inequality,

$$\frac{1}{n} s(A, B) = \frac{1}{n} \sum_{i=1}^{n} \frac{b_i}{a_i} \geqslant \left(\prod_{i=1}^{n} \frac{b_i}{a_i} \right)^{1/n} = 1,$$

with equality if and only if $b_i = a_i$ for all i.

However, in general, $\sigma(A, B) \neq \sigma(B, A)$ and σ fails the triangle inequality. Nevertheless, we will pretend that σ is a metric. For $A_1, A_2 \in X$, the σ-bisectors,

$$Bis(A_1, A_2) = \{B \in V : \sigma(A_1, B) = \sigma(A_2, B)\} = \{B \in V : tr(A_1^{-1}B) = tr(A_2^{-1}B)\}$$

are defined by an equation which is linear in the variable B. Hence, σ-bisectors are linear. Clearly,

$$\{B \in V : \sigma(A_1, B) < \sigma(A_2, B)\} \cap \{B \in V : \sigma(A_1, B) > \sigma(A_2, B)\} = \emptyset. \tag{11.1}$$

Another useful property of the 2-point invariant σ is that the function $B \mapsto \sigma(I, B)$ is proper when restricted to X. This is so because of the comparison to the invariant Finsler metric d_{max} on X:

$$\sigma(I, A) \leqslant d_{max}(I, A) \leqslant \sigma(I, A) + \log(n). \tag{11.2}$$

Here $d_{max}(I, A)$ is the logarithm of the largest eigenvalue of the matrix $A \in P$. From this, it follows that

$$\lim_{\|A\| \to \infty, A \in X} d_{max}(I, Bis(I, A) \cap X) = \infty.$$

It is also instructive to consider the invariant $s(A, B)$ in the special case $n = 2$, i.e. $G = SL(2, \mathbb{R})$, when the vector space V is 3-dimensional. Up to the harmless multiplicative factor $-1/\det(A)$, the 2-point invariant $s(A, B)$ equals $(A, B) = -tr(adj(A)B)$, where

$$adj \begin{pmatrix} a & b \\ b & c \end{pmatrix} = \begin{pmatrix} c & -b \\ -b & a \end{pmatrix}.$$

Thus, (A, B) is a bilinear form on V, still invariant under the action of G. A direct computation shows that this form is symmetric and has signature $(2, 1)$. Thus, the vector space V equipped with the form $(\cdot, \cdot)$ is a 3-dimensional Lorentzian vector space. The group G acts on V with the kernel $\{\pm I\}$, hence, through a group isomorphic to $SO^+(2, 1)$, making it the identity component of the group of all linear automorphisms of $(\cdot, \cdot)$. Let us compute the quadratic form corresponding to our bilinear form:

$$(A, A) = -tr \left\{ \begin{pmatrix} c & -b \\ -b & a \end{pmatrix} \begin{pmatrix} a & b \\ b & c \end{pmatrix} \right\} = b^2 - ac.$$

The matrix A is positive-definite if and only if $a > 0$ and $(A, A) < 0$. Hence the convex cone $P \subset V$ is a component (given by the inequality $a > 0$) of the set of time-like vectors in this Lorentzian space. The σ-bisectors in V are nothing but the Lorentzian bisectors discussed in Section 3.

With all these geometric preliminaries out of the way, we now return to the discussion of fundamental domains of discrete subgroups of G. Given a discrete subgroup $\Gamma < G$ as above, one defines the *Dirichlet–Selberg fundamental domain* of Γ in P centered at the identity matrix I:

$$DS(\Gamma, I) = \{C \in P : s(I, C) \leqslant s(\gamma^T \gamma, C), \forall \gamma \in \Gamma\}.$$

REMARK 11.1. With a minor modification, this definition generalizes to fundamental domains centered at non-identity matrices $p \in X$,

$$DS(\Gamma, p) = \{C \in P : s(p, C) \leqslant s(\gamma^T p \gamma, C), \forall \gamma \in \Gamma\}.$$

The fact that for $\gamma, \gamma' \in \Gamma$, the interiors of $\gamma DS(\Gamma, p), \gamma' DS(\Gamma, p)$ intersect if and only if $\gamma p = \gamma' p$ follows from (11.1).

LEMMA 11.2. *The collection of domains $DS(\Gamma, \gamma p), \gamma \in \Gamma$, is locally-finite in P, i.e. every compact in P intersects only finitely many cones $DS(\Gamma, \gamma p)$.*

PROOF. It suffices to prove the claim for the intersections of Dirichlet–Selberg domains with X. Suppose $R < \infty$ is such that for the R-ball $B(p,R) \subset X$ (with respect to the Finsler metric d_{max} on X) the σ-bisector $Bis(\gamma' p, \gamma p)$ bounding $D(\Gamma, \gamma p)$ intersects $B(p,R)$. Thus, there exists $x \in X$ such that

$$\sigma(\gamma p, x) \leqslant \sigma(p, x) \leqslant d_{max}(p, x) \leqslant R,$$

which (cf. (11.2)) implies that

$$d_{max}(\gamma p, x) - \log(n) \leqslant \sigma(\gamma p, x) \leqslant R.$$

In view of the proper discontinuity of the action of Γ on X, the number of such elements $\gamma \in \Gamma$ is finite. Thus, only finitely many Selberg-bisectors $Bis(\gamma' p, \gamma p)$ can intersect $B(p,R)$. □

In particular, each compact subset of P intersects only finitely many bisectors bounding DS and, hence, linearity of bisectors implies that $DS(\Gamma, I)$ is a convex polyhedral cone in P. (This cone might have infinitely many faces.) It also follows that $DS(\Gamma, p)$ satisfies all the properties of a fundamental domain of a discrete group action (cf. [**Ra**]), which justifies the name *Dirichlet–Selberg fundamental domain.*

The definition of $DS(\Gamma, p)$ suggests a slew of open questions. For instance:

QUESTION 11.3.

1. Which discrete subgroups have finitely-sided Dirichlet–Selberg domains?

2. Uniform lattices do have finitely-sided Dirichlet–Selberg domains, but what about non-uniform lattices?

3. Do Anosov subgroups[9] of G have finitely-sided Dirichlet–Selberg domains (at least for some choice of base-points p)?

In contrast to discrete subgroups of $O^+(n,1)$, it is quite unclear how *generic* are subgroups with finitely-sided Dirichlet–Selberg domains among discrete finitely generated subgroups of $SL(n,\mathbb{R})$.

By analogy with the Poincaré Fundamental Polyhedron Theorem in hyperbolic geometry (see [**EP**], and [**Ra**, Section 13.5]), one obtains a similar theorem in P, working with (relatively) closed convex polyhedral cones $C \subset P$ bounded by σ-bisectors. The paper [**EP**] is especially useful here, since it focuses on algorithmic aspects of the Poincaré Fundamental Polyhedron Theorem in the BSS computability model. Below is a review of the formulation of this theorem, adopted to the setting of domains bounded by σ-bisectors.

Suppose that C is a finitely-sided convex polyhedral cone in P, bounded by σ-bisectors of the form $Bis(p, \gamma_i p)$, where p is a chosen point in X and $\gamma_1, \ldots, \gamma_q$ are certain elements of G. We further assume that the facets in C are matched in pairs F_i, F_i' by the elements $\gamma_i \in G$, $i = 1, \ldots, q$, so that $\gamma_i : F_i \to F_i'$, $\gamma_i^{-1} : F_i' \to F_i$, and

$$F_i' \subset Bis(p, \gamma_i p), \quad F_i \subset Bis(p, \gamma_i^{-1} p), i = 1, \ldots, q.$$

Then, similarly to the case of the hyperbolic space, one defines *ridge-cycles* in ∂C corresponding to the *ridges*, which are codimension 2 faces E of C. Every such ridge-cycle is a finite sequence of elements $\gamma_i^{\pm 1}$,

$$(\gamma_{i_1}^{\pm 1}, \ldots, \gamma_{i_\ell}^{\pm 1}).$$

[9] See section 12.

Specifically, assuming that the ridge E is the intersection of, say, facets F_{i_1}, F_{i_ℓ}, one starts the cycle with the generator γ_{i_1} pairing F_{i_1} and F'_{i_1}. The image $\gamma_{i_1}(E)$ is a ridge of the facet F'_{i_1}, hence, is the intersection of F'_{i_1} and F_{i_2}. Then the second element in the cycle is the generator γ_{i_2} pairing F_{i_2} and F'_{i_2}, etc. Since the cone C has only finitely many faces, eventually, we come back to the original ridge E, completing the cycle. This yields the product

$$\beta_{E,\ell} = \gamma_{i_\ell}^{\pm 1} \circ \cdots \circ \gamma_{i_1}^{\pm 1}.$$

Ridge-cycles are defined so that ℓ is the least natural number such that $\beta_{E,\ell}(E) = E$. The conditions of the Poincaré Fundamental Polyhedron Theorem require that each β_ℓ has finite order N_E and fixes E pointwise. Moreover, let U_s denote a small neighborhood of the ridge $\beta_{E,s}(E)$ in the cone C. Set

$$F_E := \bigcup_{s=1}^{\ell-1} \beta_{E,s}(U_s).$$

Then the ridge-cycle condition also requires F_E together with its images

$$\beta_{E,\ell}(F_E), \beta_{E,k}^2(F_E), \ldots, \beta_{E,k}^{N_E-1}(F_E)$$

to form a perfect tiling of a neighborhood of E in P. This tiling condition can be reformulated in terms of Riemannian angles. Pick a point $x = x_0 \in E$. For each $s = 0, 1, \ldots, \ell - 1$, we set $x_s = \beta_{E,s}(x)$. Let $\alpha_s(x)$ denote the Riemannian (with respect to the G-invariant Riemannian metric on X) dihedral angle between the facets of $C \cap X$ at $x_s \in \beta_{E,s}(E) \cap X$. Then one requires

$$\alpha_E := \sum_{s=1}^{\ell} \alpha_s = \frac{2\pi}{N_E}. \tag{11.3}$$

(If this holds for one choice of x, then it holds for all choices.)

REMARK 11.4. Yukun Du, [**D**], recently defined (for generic generators γ_i) non-Riemannian analogues of angles between the above bisectors, which are G-invariant, satisfy the natural additivity property and also the property that the *neighborhood tiling* above is equivalent to the angle-sum condition (11.3). These "angles" do not depend on choices of points $x \in E$ and are defined in terms of linear algebra. Hence, they are more amenable to computations than the Riemannian angles.

A polyhedral cone C as above is a *pre-Dirichlet–Selberg domain* for the subgroup $\Gamma < G$ generated by $\gamma_1, \ldots, \gamma_q$: If Γ is discrete (which is, a priori, unclear), its actual Dirichlet–Selberg domain $DS(\Gamma, p)$ is contained in the cone C.

The last condition of the Poincaré Fundamental Polyhedron Theorem is the least pleasant one (it is void if $D = C \cap X$ is compact). The face-pairing transformations γ_i of $D := C \cap X$ define an equivalence relation $\sim$ on D generated by

$$x \sim \gamma_i(x), x \in F_i \cap X.$$

The ridge-cycle conditions above imply that the quotient-space $D/\sim$ has natural structure of a Riemannian orbifold modeled on the symmetric space X. Then the last condition requires this orbifold to be metrically complete. In the setting of the hyperbolic space, this metric completeness requirement can be replaced by a more computable "ideal vertex cycle" condition that can be found in [**EP**], [**Ra**] (see Theorems 13.4.5 and Theorems 13.4.7 in Ratcliffe's book). I currently do not know how to formulate a similar condition in the setting of convex cones in P as above. Below, is a review of the "ideal vertex cycle" condition in the context of hyperbolic spaces. Following [**Ra**], a *cusp point* of a finitely-sided convex polyhedron $Q \subset \mathbb{H}^n$ is a point v of the closure of Q in $\overline{\mathbb{H}^n} = \mathbb{H}^n \cup S^{n-1}$ (here we use the Klein model of the hyperbolic space) which equals to the intersection of closures in $\overline{\mathbb{H}^n}$ of all faces of Q whose closures contain v. Then, similarly, to the ridge-cycles, one defines ideal vertex cycles of such cusp points v. Let ℓ be the least integer such that the product $\beta_{v,\ell}$ sends v to itself. The *ideal vertex cycle* condition then is that every such $\beta_{v,\ell}$ is either elliptic or parabolic.

Lastly, returning to the pre-Dirichlet–Selberg domains C, one has:

THEOREM 11.4. *The above ridge-cycles and completeness conditions are necessary and sufficient for C to be a fundamental domain for the action on P of the subgroup $\Gamma < G$ generated by the elements $\gamma_1, \gamma_2, \dots$. Moreover, Γ has the presentation in the above generating set, where the relators are the products $\beta_{E,\ell}$ of the ridge-cycles, and E runs through the set of equivalence classes of the ridges in C.*

The proof of this theorem is exactly the same as in the hyperbolic case, see [**EP, Ra**]. The hardest part in a computational implementation of this theorem is the completeness condition and presently, we do not know how to deal with the issue (see, however, Conjecture 11.5 below). However, in the case of a compact fundamental domain the completeness is automatic and one obtains an algorithm for computing a fundamental domain and, hence, a finite presentation of a uniform lattice in G. Let $\Gamma < SL(n,\mathbb{R})$ be a uniform arithmetic lattice, given by its arithmetic data.

(1) For each $N \in \mathbb{N}$, compute the subset Γ_N matrices $A \in \Gamma$ such that $s(I, A) \leqslant N$, which is a finite search.
(2) Compute the intersection C_N of closed half-spaces $\{s(I,x) \leqslant s(A^TA, x)\}$ (in the vector space V of symmetric matrices) for $A \in \Gamma_N$. The intersection $C_N \cap P$ will be called a *partial Dirichlet–Selberg domain* of Γ.
(3) Check if this intersection is contained in the cone P of positive-definite matrices.
(4) If it is not, then increase N to $N+1$ and repeat.
(5) Suppose, yes, then compute the *Selberg-radius* $\delta = \max\{\sigma(I,x) : x \in D_N\}$ of $D_N := C_N \cap X$ (imposing the extra condition det $= 1$).
(6) Next, find all $A \in \Gamma$ such that $N \leqslant \sigma(I, A^TA) \leqslant 2\delta + 2\log(n)$.
(7) $D_{2\delta+2\log(n)}$ will be the Dirichlet–Selberg domain of Γ.

The reader can find examples (my list is far from exhaustive) of similar algorithms in [**R1, R2**] (in the case of subgroups of $PSL(2,\mathbb{C})$ and $PSL(2,\mathbb{R})$ respectively), in [**M**] (in the case of subgroups of $PSL(2,\mathbb{R})$), in [**Li, P, Si**] (in the case of subgroups of $PSL(2,\mathbb{C})$), in [**C**] (in the case of groups acting on the bidisk), [**CS**] (in the case of groups acting on the complex 2-ball) and [**EP**] for subgroups of $O(n,1)$. See also [**J**] for a discussion of Siegel fundamental domains for the subgroup $Sp(2n,\mathbb{Z}) < Sp(2n,\mathbb{R})$.

Lastly, for general discrete subgroups $\Gamma < G$, we have:

CONJECTURE 11.5. *Each pre-Dirichlet–Selberg domain of Γ (regardless of its compactness!) satisfies the completeness condition and, hence, is a fundamental domain of Γ.*

One reason to be optimistic is that the analogous statement does hold for subgroups of $O(n,1)$ and pre-Dirichlet domains of such subgroups. The key reason is that if Q is a pre-Dirichlet domain in $\mathbb{H}^n$ centered at a point $p \in \mathbb{H}^n$, then for each point $v \in S^{n-1}$ in the closure of $Bis(p, \gamma_i^{\pm 1} p)$, the points $p, \gamma_i^{\pm 1}(p)$ lie on the same *horosphere* in $\mathbb{H}^n$ based at v. For each cusp-point v of Q we take the horosphere Σ_v based at v and passing through p. Suppose that

$$(\gamma_{i_1}^{\pm 1}, \ldots, \gamma_{i_\ell}^{\pm 1})$$

is an ideal vertex cycle of a cusp point v of Q and

$$\gamma_{i_1}^{\pm 1}(v) = v_1, \gamma_{i_2}^{\pm 1}(v_1) = v_2, \ldots, \gamma_{i_\ell}^{\pm 1}(v_{\ell-1}) = v.$$

The above observation about boundary points of bisectors then implies that

$$\gamma_{i_1}^{\pm 1} : \Sigma_v \to \Sigma_{v_1}, \quad \gamma_{i_2}^{\pm 1} : \Sigma_{v_1} \to \Sigma_{v_2}, \ldots, \quad \gamma_{i_\ell}^{\pm 1} : \Sigma_{v_{\ell-1}} \to \Sigma_v.$$

Applying this to the product

$$\beta_{v,\ell} = \gamma_{i_\ell}^{\pm 1} \cdots \circ \gamma_{i_1}^{\pm 1},$$

we conclude that it sends the horosphere Σ_v back to itself. Hence, $\beta_{v,\ell}$ is elliptic or parabolic, verifying the ideal vertex cycle condition.

12. Computational aspects of Anosov subgroups

We will again limit the discussion to the case of discrete subgroups of $G = SL(n, \mathbb{R})$. While general *undistorted* finitely generated subgroups of G are rather poorly-behaved (for instance, they need not be finitely presentable), the *Anosov* condition below eliminates various pathologies and results in a class of subgroups which share many desirable properties with undistorted subgroups of $O^+(n,1)$. The *Anosov property* was originally formulated for discrete subgroups Γ (of semisimple Lie groups) by Labourie, Guichard and Wienhard, see [**La**], [**GW**]. While it is defined *relative* to a certain parabolic subgroup P of G, for the sake of simplicity, I limit myself to the discussion when P is a minimal parabolic subgroup (i.e. Borel subgroup) of $G = SL(n, \mathbb{R})$, i.e. the subgroup of upper triangular matrices. To simplify the terminology, we will refer to such subgroups simply as *Anosov*. The following is a treatment of Anosov subgroups following our work with Leeb and Porti, [**KLP2**] and [**KL**].

First, let us revisit the notion of discreteness for subgroups of $G = SL(n, \mathbb{R})$: A subgroup $\Gamma < G$ is discrete if every sequence of distinct matrices γ_i in Γ diverges to infinity, $\|\gamma_i\| \to \infty$. Equivalently, the sequence of highest singular values of γ_i's diverges to infinity.

Looking at the singular values of these matrices arranged in the decreasing order,

$$\sigma_1(\gamma_i) \geqslant \sigma_2(\gamma_i) \geqslant \cdots \geqslant \sigma_n(\gamma_i),$$

and their asymptotics as $i \to \infty$, one realizes that this divergence to infinity can happen in quantitatively different ways. For instance, a sequence of matrices is called *regular* if each sequence of successive quotients $\frac{\sigma_k(\gamma_i)}{\sigma_{k+1}(\gamma_i)}$ diverges to infinity. Accordingly, *regularity* of a discrete subgroup Γ means that every unbounded sequence in it is regular. (The regularity condition can be weakened to *partial regularity* by looking at the ratios of *some* of the successive singular values, leading to an interesting theory as well: This corresponds to the notion of P-Anosov subgroups for general parabolic subgroups $P < G$.) Regularity is equivalent to discreteness if $n = 2$ but not for $n \geqslant 3$. For instance, the subgroup of matrices with integer coefficients, $SL(n, \mathbb{Z})$, is discrete but is not even partially regular if $n \geqslant 3$. (One way to see this lack of regularity is to observe that $SL(n, \mathbb{Z})$ contains diagonalizable subgroups isomorphic to $\mathbb{Z}^{n-1}$ and such subgroups are easily seen to be non-regular with respect to any parabolic subgroup $P < G$.)

So far, our discussion was in terms of linear algebra; in order to get the actual Anosov condition, one connects linear algebra with the geometry of Γ itself, assuming that Γ is finitely generated, equipped with a word-metric d_Γ. It is not hard to see that the ratios of singular values as above cannot diverge to infinity at rate faster than exponential with respect to $d_\Gamma(1, \gamma_i)$, but they *can* diverge to infinity subexponentially, even linearly. (This happens, for instance, in the case of $SL(2, \mathbb{Z})$ when we consider the sequence of powers of a unipotent matrix.) This observation leads to a definition, which (in a more geometric form) first appeared in our work with Leeb and Porti:

DEFINITION 12.1. A (discrete) finitely generated subgroup $\Gamma < G = SL(n, \mathbb{R})$ is called *URU* if there exists $A > 0$ such that for every $\gamma \in \Gamma$,

$$\frac{\sigma_k(\gamma)}{\sigma_{k+1}(\gamma)} \geqslant A^{-1} \exp(A \cdot d_\Gamma(1, \gamma)), k = 1, \ldots, n-1.$$

In particular, this definition includes the property that Γ is undistorted in G and is a regular subgroup. However, the regularity condition appearing in this definition is a bit stronger than the one formulated above, it is called *uniform regularity* in [**KLP1, KLP2**]. We will not define it here (as it will not be needed), but only note that URU stands for *uniformly regular undistorted.* It is proven in [**KLP2**] that every URU subgroup is word-hyperbolic.

Given a hyperbolic group Γ and a homomorphism $\rho : \Gamma \to G$, one says that ρ *is Anosov* if ρ has finite kernel and Anosov image.

It was proven in [**KLP1**] that the Anosov property for group homomorphisms is semidecidable. The KLP algorithm for testing the Anosov property is very similar to the one described in Section 9 for representations to $O^+(n, 1)$ using the *midpoint test.* The main differences with the hyperbolic case are:

(1) One adds a regularity test for the geodesic segments $m_j m_{j+1}$ connecting the midpoints of the geodesic segments $\rho(w_j)(x)\rho(w_{j+1})(x)$ in the space X of positive definite matrices with unit determinant. (Here w_j, w_{j+1} are the words appearing in geodesic quadruples in the midpoint test described in Section 9.) For the geodesic segment connecting the identity matrix to a matrix $m \in X$, the regularity condition amounts to checking that m satisfies the eigenvalue inequalities

$$\frac{\lambda_k(m)}{\lambda_{k+1}(m)} \geqslant r > 1, \quad k = 1, 2, \ldots, n-1.$$

(2) The Riemannian angles α appearing in Section 9 are replaced with certain ζ-angles, which I will discuss below.
(3) In the analogue of the inequality in Theorem 6.2, the distances L and ζ-angles α are *decoupled*: One requires that $L \geqslant L_n(r)$ and $\alpha \geqslant \pi - \epsilon_n(r)$, where $L_n(r)$ and $\epsilon_n(r)$ are certain functions.

In [**KLP1**] the existence of the functions $L_n = L_n(r)$ and $\epsilon_n = \epsilon_n(r)$ (for which the algorithm works) was established by certain continuity arguments. Max

Riestenberg in his PhD thesis, [**Ri**], computed these functions explicitly, making it, in theory, possible, to test if the given representation is Anosov, in particular, has finite kernel and discrete image. The KLP algorithm then runs essentially as in Section 9, except, in addition to increasing N, one also decreases $r > 1$, taking it equal, say, to $1 + \frac{1}{N}$. The algorithm terminates if and only if ρ is Anosov.

Below is a definition of ζ-angles adapted to the setting of the symmetric space X of the group $SL(n, \mathbb{R})$. We first define ζ-angles between tangent directions $\mathbf{u}, \mathbf{v}$, i.e. nonzero vectors in the tangent space $T_I X$ at the identity matrix $I \in X$. This tangent space is nothing but the space of traceless symmetric matrices. The ζ-angle is defined only between *regular* matrices $\mathbf{u}, \mathbf{v} \in T_I X$, meaning that the eigenvalues of $\mathbf{u}$ and of $\mathbf{v}$ are pairwise distinct.

The ζ-angle is defined with respect to a fixed diagonal matrix

$$\zeta = Diag(\zeta_1, \ldots, \zeta_n)$$

satisfying the following conditions:

(1) $\zeta_i = -\zeta_{n-i}$, $i = 1, 2, \ldots$
(2) $\sum_{i=1}^n \zeta_i = 0$.
(3) $\zeta_i > \zeta_{i+1}$ for all $i \leqslant n - 1$.

There is no canonical choice of such vectors, one can take, for instance, ζ coming from the *sum of positive coroots* of the root system of type A, namely,

$$\zeta_1 = (n-1), \quad \zeta_{i+1} = \zeta_i - 2, i = 1, 2, \ldots, n-1.$$

In particular (with this choice), if n is odd, then (for $n - 1 = 2k$),

$$\|\zeta\|^2 = 2 \sum_{i=1}^{k} (2i)^2,$$

while if $n = 2k$ is even, then

$$\|\zeta\|^2 = 2 \sum_{i=1}^{k} (2i-1)^2.$$

The most important property that ζ has, is that it belongs to the *open Weyl chamber* defined by the inequalities

$$\zeta_1 > \zeta_2 > \cdots > \zeta_n.$$

Next, given a matrix $\mathbf{u} \in T_I X$ with the eigenvalues $\lambda_1 > \lambda_2 > \cdots > \lambda_n$, let $Q \in O(n)$ be the matrix which diagonalizes $\mathbf{u}$, i.e.

$$\mathbf{u} = Q^T Diag(\lambda_1, \ldots, \lambda_n) Q.$$

Then define the *ζ-direction* of $\mathbf{u}$ as

$$\mathbf{u}_\zeta := Q^T \zeta Q.$$

Recall that $T_I X$ is equipped with the Riemannian metric $\langle \mathbf{x}, \mathbf{y} \rangle = tr(\mathbf{x}\mathbf{y})$. The ζ-angle between $\mathbf{u}, \mathbf{v}$ is defined as the Riemannian angle between the directions $\mathbf{u}_\zeta$ and $\mathbf{v}_\zeta$. In other words,

$$\cos(\angle_I^\zeta(\mathbf{u}, \mathbf{v})) = \frac{tr(\mathbf{u}_\zeta \mathbf{v}_\zeta)}{\|\zeta\|^2}.$$

By the construction, such angles are invariant under the action of $O(n)$ on the tangent space $T_I X$.

We next define ζ-angles for triangles ΔIM_1M_2 in X, by measuring the angle at the corner I of the triangle. Define matrices $\mathbf{m}_i \in T_IX$ by

$$\mathbf{m}_i = \log(M_i), i = 1, 2.$$

Then set

$$\angle_I^\zeta(M_1, M_2) := \angle_I^\zeta(\mathbf{m}_1, \mathbf{m}_2) = \angle_I^\zeta(\log M_1, \log M_2).$$

Lastly, in order to define ζ-angles of general triangles ΔMM_1M_2 in X, we impose the G-invariance of such angles. Suppose that $M = g^T g, g \in G$. Then set

$$\angle_M^\zeta(M_1, M_2) := \angle_I^\zeta((g^{-1})^T M_1 g^{-1}, (g^{-1})^T M_2 g^{-1}).$$

We will not define the the functions $L_n = L_n(r)$ and $\epsilon_n = \epsilon_n(r)$ and refer instead to the work of Max Riestenberg, [**Ri**]. Note that the computational feasibility of the KLP algorithm is currently unclear even in the case of subgroups of $O(n, 1)$, which should be addressed first (before the case of discrete subgroups of $SL(n, \mathbb{R})$ is discussed). To the best of our knowledge, this was never done.

Acknowledgment

I am grateful to the referees of the paper for doing a very thorough refereeing job.

References

[Bo1] B. H. Bowditch, *Geometrical finiteness for hyperbolic groups*, J. Funct. Anal. **113** (1993), no. 2, 245–317, DOI 10.1006/jfan.1993.1052. MR1218098

[Bo2] B. H. Bowditch, *Geometrical finiteness with variable negative curvature*, Duke Math. J. **77** (1995), no. 1, 229–274, DOI 10.1215/S0012-7094-95-07709-6. MR1317633

[C] Harvey Cohn, *Some computer-assisted topological models of Hilbert fundamental domains*, Math. Comp. **23** (1969), 475–487, DOI 10.2307/2004375. MR246820

[CS] D. I. Cartwright and T. Steger, *Finding generators and relations for groups acting on the hyperbolic ball*, Preprint, 2017. `arXiv:1701.02452`. Software documentation: `https://www.maths.usyd.edu.au/u/donaldc/fakeprojectiveplanes/`

[DK] Cornelia Druţu and Michael Kapovich, *Geometric group theory*, American Mathematical Society Colloquium Publications, vol. 63, American Mathematical Society, Providence, RI, 2018. With an appendix by Bogdan Nica, DOI 10.1090/coll/063. MR3753580

[D] Y. Du, *Hyperplanes in the symmetric space $SL(n, \mathbb{R})/SO(n)$*, Preprint.

[EH] David B. A. Epstein and Derek F. Holt, *Computation in word-hyperbolic groups*, Internat. J. Algebra Comput. **11** (2001), no. 4, 467–487, DOI 10.1142/S0218196701000619. MR1850213

[EP] David B. A. Epstein and Carlo Petronio, *An exposition of Poincaré's polyhedron theorem*, Enseign. Math. (2) **40** (1994), no. 1-2, 113–170. MR1279064

[FR] Elena Fuchs and Igor Rivin, *Generic thinness in finitely generated subgroups of* $\mathrm{SL}_n(\mathbb{Z})$, Int. Math. Res. Not. IMRN **17** (2017), 5385–5414, DOI 10.1093/imrn/rnw136. MR3694603

[G1] Jane Gilman, *Two-generator discrete subgroups of* $\mathrm{PSL}(2, \mathbf{R})$, Mem. Amer. Math. Soc. **117** (1995), no. 561, x+204, DOI 10.1090/memo/0561. MR1290281

[G2] Jane Gilman, *Algorithms, complexity and discreteness criteria in* $\mathrm{PSL}(2, \mathbf{C})$, J. Anal. Math. **73** (1997), 91–114, DOI 10.1007/BF02788139. MR1616469

[G3] J. Gilman, *Computability models: Algebraic, topological and geometric algorithms*, Computational Aspects of Discrete Subgroups of Lie Groups (Alla Detinko, Michael Kapovich, Alex Kontorovich, Peter Sarnak, and Richard Schwartz, eds.), Contemporary Mathematics, vol. 783, Amer. Math. Soc.,Providence, RI, 2023.

[GM] J. Gilman and B. Maskit, *An algorithm for 2-generator Fuchsian groups*, Michigan Math. J. **38** (1991), no. 1, 13–32, DOI 10.1307/mmj/1029004258. MR1091506

[GW] Olivier Guichard and Anna Wienhard, *Anosov representations: domains of discontinuity and applications*, Invent. Math. **190** (2012), no. 2, 357–438, DOI 10.1007/s00222-012-0382-7. MR2981818

[J] C. Jaber, *Algorithmic approaches to Siegel's fundamental domain*, PhD Thesis, Université de Bourgogne, 2017. `https://tel.archives-ouvertes.fr/tel-01813184/document`

[K1] Michael Kapovich, *Discreteness is undecidable*, Internat. J. Algebra Comput. **26** (2016), no. 3, 467–472, DOI 10.1142/S0218196716500193. MR3506344

[K2] M. Kapovich, *A note on properly discontinuous actions*, Preprint, 2022. `https://www.math.ucdavis.edu/~kapovich/EPR/prop-disc.pdf`

[KL] Michael Kapovich and Bernhard Leeb, *Discrete isometry groups of symmetric spaces*, Handbook of group actions. Vol. IV, Adv. Lect. Math. (ALM), vol. 41, Int. Press, Somerville, MA, 2018, pp. 191–290. MR3888689

[KLP1] M. Kapovich, B. Leeb and J. Porti, *Morse actions of discrete groups on symmetric spaces*, arXiv e-print, 2014. `arXiv:1403.7671`.

[KLP2] Michael Kapovich, Bernhard Leeb, and Joan Porti, *A Morse lemma for quasigeodesics in symmetric spaces and Euclidean buildings*, Geom. Topol. **22** (2018), no. 7, 3827–3923, DOI 10.2140/gt.2018.22.3827. MR3890767

[KLi] Michael Kapovich and Beibei Liu, *Geometric finiteness in negatively pinched Hadamard manifolds*, Ann. Acad. Sci. Fenn. Math. **44** (2019), no. 2, 841–875, DOI 10.5186/aasfm.2019.4444. MR3973544

[La] François Labourie, *Anosov flows, surface groups and curves in projective space*, Invent. Math. **165** (2006), no. 1, 51–114, DOI 10.1007/s00222-005-0487-3. MR2221137

[Li] M. Lipyanskiy, *Computer-assisted application of Poincare's fundamental polyhedron theorem*, Preprint, 2001, `http://www.math.columbia.edu/~ums/pdf/poincare.pdf`.

[M] Gregory Muller, *Computing a generating set of arithmetic Kleinian groups*, Teichmüller theory and moduli problem, Ramanujan Math. Soc. Lect. Notes Ser., vol. 10, Ramanujan Math. Soc., Mysore, 2010, pp. 513–517. MR2667570

[P] Aurel Page, *Computing arithmetic Kleinian groups*, Math. Comp. **84** (2015), no. 295, 2361–2390, DOI 10.1090/S0025-5718-2015-02939-1. MR3356030

[Ra] John G. Ratcliffe, *Foundations of hyperbolic manifolds*, Graduate Texts in Mathematics, vol. 149, Springer, Cham, [2019] ©2019. Third edition [of 1299730], DOI 10.1007/978-3-030-31597-9. MR4221225

[Ri] J. M. Riestenberg, *A quantified local-to-global principle for Morse quasigeodesics*, Preprint, 2021. `arXiv:2101.07162`.

[R1] Robert Riley, *Applications of a computer implementation of Poincaré's theorem on fundamental polyhedra*, Math. Comp. **40** (1983), no. 162, 607–632, DOI 10.2307/2007537. MR689477

[R2] Robert Riley, *Nielsen's algorithm to decide whether a group is Fuchsian*, In the tradition of Ahlfors and Bers (Stony Brook, NY, 1998), Contemp. Math., vol. 256, Amer. Math. Soc., Providence, RI, 2000, pp. 255–270, DOI 10.1090/conm/256/04012. MR1759685

[Se] Atle Selberg, *On discontinuous groups in higher-dimensional symmetric spaces*, Contributions to function theory (Internat. Colloq. Function Theory, Bombay, 1960), Tata Institute of Fundamental Research, Bombay, 1960, pp. 147–164. MR0130324

[Si] K. Siegel, *Applying Poincare's polyhedron theorem to groups of hyperbolic isometries*, Preprint, `https://kylersiegel.xyz/Poincare.pdf`.

Department of Mathematics, University of California, Davis, California 95616
Email address: `kapovich@ucdavis.edu`

Contemporary Mathematics
Volume **783**, 2023
https://doi.org/10.1090/conm/783/15737

List of problems on discrete subgroups of Lie groups and their computational aspects

Michael Kapovich, Alla Detinko, and Alex Kontorovich

Abstract. In this paper we present a problem list pertaining to discrete subgroups of Lie groups and their computational aspects, consisting mostly of the problems collected during the ICERM workshop "Computational Aspects of Discrete Subgroups of Lie Groups" held in June of 2021.

In this paper we present a problem list, consisting mostly of the problems collected during the ICERM workshop held in June of 2021. However, some of the problems are older, some of these go back to the 1970s. Many of the problems are purely theoretical, while some have an obvious computational flavor.

1. Background

In this section we collect definitions and basic facts about abstract groups and discrete subgroups of Lie groups that are used in what follows.

Group theory. We begin with a discussion of some group-theoretic notions. Most of these notions deal with the *subgroup structure* of abstract groups.

An abstract group G is said to satisfy a property P *virtually* if there exists a finite-index subgroup of G which satisfies P.

An abstract group G is said to be *a surface group* if it is isomorphic to the fundamental group of a closed (i.e. compact with empty boundary) surface of negative Euler characteristic.

An abstract group G is said to be *coherent* if every finitely generated subgroup of G is also finitely-presentable.

A subgroup H of a group G is called *maximal* if there is no proper subgroup between H and G. Some maximal subgroups have finite index in G (for instance, subgroups of prime index are always maximal). Of interest to us are maximal subgroups of infinite index; we will refer to these as *strictly maximal*.

A group G is said to satisfy the *Howson property* if the intersection of any two finitely generated subgroups is again finitely generated.

2020 *Mathematics Subject Classification.* Primary 22E40, 20-08; Secondary 20F67, 53C35.

The third author was supported by an NSF CAREER grant DMS-1455705, an NSF FRG grant DMS-1463940, NSF grant DMS-1802119, a BSF grant number 2014099, and the Simons Foundation through MoMath's Distinguished Visiting Professorship for the Public Dissemination of Mathematics.

The property is named after A. G. Howson, who proved in [**31**] that free groups satisfy this property. In contrast, if F_r is the free group of rank $r \geq 2$, then $F_r \times \mathbb{Z}$ does not satisfy the Howson property (see Example 1.1 below). In particular, $SL(n, \mathbb{Z}), n \geq 4$, does not satisfy the Howson property either (since it always contains $F_r \times \mathbb{Z}$). On the other hand, all discrete subgroups of $PSL(2, \mathbb{R})$ satisfy the Howson property (see e.g. [**27**] for surface groups). More generally, every finitely generated discrete subgroup of $PSL(2, \mathbb{C})$ which is not a lattice satisfies the Howson property; see e.g. [**30**].

Even more generally, if Γ_1, Γ_2 are geometrically finite subgroups of a discrete subgroup Γ in a rank 1 Lie group (see below), then the intersection $\Gamma_1 \cap \Gamma_2$ is again geometrically finite, hence, finitely generated. A proof of this result again appears in Hempel's paper [**30**]: While he only works with discrete subgroups of $PSL(2, \mathbb{C})$, his proof is also valid for subgroups of other rank 1 Lie groups. In contrast, the Howson property fails *for all* lattices in $PSL(2, \mathbb{C})$: It was noted by Hempel in [**30**] that the property fails for the fundamental groups of 3-dimensional manifolds fibering over the circle. Due to the work of Agol and Wise, it is known that all finite volume hyperbolic 3-manifolds admit finite-sheeted covering spaces which fiber over the circle.

Distortion. Let G, H be finitely generated groups equipped with word metrics d_G, d_H respectively. Assume that H is a subgroup of G. Then the *distortion function* $\delta(n)$ for the inclusion map $H \to G$ is defined as follows:

$$\delta(n) = \max\{d_H(1, h) : h \in H, d_G(1, h) \leq n\}.$$

For instance, $\delta(n)$ is linear if and only if the inclusion $H \to G$ is *bi-Lipschitz*, i.e. there exists a constant A such that

$$d_H(1, h) \leq A d_G(1, h)$$

for all $h \in H$. Subgroups with linear distortion are said to be *undistorted*. The same concept applies in the case of an isometric group action $H \times X \to X$ of a finitely generated group H on a metric space X. The distortion function (relative to a point $x \in X$) of this action is

$$\delta_x(n) = \sup\{d_H(1, h) : h \in H, d_X(x, hx) \leq n\}.$$

Geometric finiteness. We now turn to the discussion of discrete subgroups of Lie groups.

Many problems which are open for higher rank lattices (and their subgroups) are well-understood in the case of discrete subgroups of rank 1 Lie groups G, or, at least, for *geometrically finite* subgroups of G. Hence, we begin by reviewing some elements of the theory of discrete subgroups of rank 1 Lie groups.

Geometric finiteness. Let G be a rank 1 Lie group (with finite center and finitely many connected components). Then the symmetric space corresponding to G is the quotient $X = G/K$, where $K < G$ is a maximal compact subgroup. One equips X with the projection of a Riemannian metric on G which is K-right-invariant and G-left-invariant. The Riemannian manifold X is then complete, simply-connected and has sectional curvature in some interval $[-b, -a]$, where $a > 0$. A subgroup $\Gamma < G$ is discrete if and only if it acts properly discontinuously on X.

There are two most tractable classes of discrete subgroups $\Gamma < G$: *Convex-cocompact* and, more generally, *geometrically finite*.

DEFINITION 1.1. A discrete subgroup $\Gamma < G$ is said to be *convex-cocompact* if there exists a nonempty closed convex Γ-invariant subset $C \subset X$ such that C/Γ is compact.

Every convex-cocompact subgroup is finitely-presentable and, moreover, is Gromov-hyperbolic. Examples of convex-cocompact subgroups are given, for instance, by uniform lattices in Γ. We refer the reader to the surveys [**32**, **34**] for other interesting examples.

Here is a useful criterion of convex cocompactness: A subgroup $\Gamma < G$ is convex-cocompact if and only if the following two properties hold (see [**12**]):

(a) Γ is finitely generated. We let d_Γ denote the word metric on Γ with respect to some finite generating set.

(b) For one (equivalently, every) $x \in X$, the *orbit map*

$$o_x : \Gamma \to \Gamma x \subset X, \quad o_x(\gamma) = \gamma x$$

is a *quasi-isometric embedding* $(\Gamma, d_\Gamma) \to X$, where X is equipped with its Riemannian distance function d_X. In the case at hand, the map o_x is a quasi-isometric embedding if and only if there exists a constant L such that for each $\gamma \in \Gamma$

$$L^{-1} d_\Gamma(\gamma, 1_\Gamma) - L \leq d_X(x, \gamma x) = d_X(x, o_x(\gamma)).$$

One also says that such subgroups $\Gamma < G$ are *undistorted* (the action of Γ on X is undistorted).

DEFINITION 1.2. A subgroup $\Gamma < G$ is said to be *geometrically finite* if the following two conditions hold:

(a) There exists a nonempty closed convex subset $C \subset X$ such that C/Γ has finite and positive volume.

(b) Orders of finite-order elements in Γ are bounded from above.

Note that, in view of Selberg's lemma, the second condition is automatically satisfied if Γ is finitely generated. Examples of geometrically finite subgroups of G are given by lattices in G, in which case $C = X$. While geometrically finite subgroups of G are, in general, not undistorted, the distortion of the word metric of Γ with respect to the metric d_X is *at worst exponential*: There exists a constant A such that

$$\log_A(d_\Gamma(\gamma, 1_\Gamma)) - A \leq d_X(x, \gamma x) = d_X(x, o_x(\gamma))$$

for all $\gamma \in \Gamma$.

Discrete subgroups of higher rank Lie groups. A subgroup in a lattice Γ in an algebraic group G is called *thin* if it is Zariski dense but has infinite index in Γ.

An element of $SL(n, \mathbb{R})$ that is diagonalizable over $\mathbb{R}$ is said to be *regular* if it has distinct eigenvalues; it is called *singular* otherwise. A rank 2 free abelian subgroup of $SL(n, \mathbb{R})$ is said to be *supersingular* if it is generated by two singular elements, whose product is also singular. More generally, an element of a Lie group with finitely many connected components is called regular if its image under the adjoint representation is regular. In the theory of P-Anosov subgroups $\Gamma < G$ (which we will briefly discuss in a moment) one also meets a *relative* notion of regularity, relative to the parabolic group $P < G$. In particular, all infinite order elements of P-Anosov subgroups $\Gamma < G$ are P-regular.

Currently, there is no clarity on what higher-rank analogues of convex-cocompactness and geometric finiteness should be, i.e. generalizations of these rank 1 notions to discrete subgroups Γ of semisimple Lie groups G (with finitely many components and finite center), such that the real rank of G is ≥ 2. One of the generalizations of the class of convex-cocompact subgroups is given by *P-Anosov subgroups*, where P is a parabolic subgroup of G. We refer the reader to the paper [**35**, section 11] in this volume for some discussion of these and references. There is even less clarity regarding geometric finiteness; initial steps in this direction are taken in [**36**], where various relativizations of Anosov subgroups are proposed and relations between them are established. However, none of these classes contains any lattices in higher rank Lie groups. Another approach to generalizing convex-cocompactness in higher rank appears in [**18**].

The next example shows that the Howson property fails in higher rank, even for intersections of Anosov subgroups of lattices.

EXAMPLE 1.1. There exists a discrete subgroup $\Gamma < SL(3,\mathbb{R})$ isomorphic to $F_2 \times \mathbb{Z}$ which contains two Anosov subgroups whose intersection is not finitely generated.

To find such a subgroup, consider the standard embedding $SO(2,1) < SL(3,\mathbb{R})$ and note that it commutes with a subgroup C (isomorphic to $\mathbb{R}$) consisting of singular matrices. Let $\Gamma_1 < SO(2,1)$ be a Schottky subgroup isomorphic to the rank 2 free group F_2 and generated by elements a, b. Let c be a non-trivial element of C. The subgroup Γ generated by a, b and c is discrete, and isomorphic to $F_2 \times \mathbb{Z}$. Define $\Gamma_2 < \Gamma$ to be the subgroup generated by a and the product bc. Then the intersection $\Gamma_1 \cap \Gamma_2$ is the normal closure of $\langle a \rangle$ in F_2 (see [**51**]), hence, it is not finitely generated (see also an explanation in [**30**]).

At the same time, Γ_1 is an Anosov subgroup of the rank 1 Lie group $SO(2,1)$, hence, it is an Anosov subgroup in $SL(3,\mathbb{R})$ (see e.g. [**28**]). With a bit more work, it follows that Γ_2 is also Anosov. For instance, if c is sufficiently close to $1 \in SL(3,\mathbb{R})$, then the Anosov property of Γ_2 follows from the stability of Anosov subgroups; see again [**28**] or [**37**].

2. $SL(2,\mathbb{Z})$-related problems

Take the congruence subgroup $\Gamma(2) < SL(2,\mathbb{Z})$ and let Λ denote the commutator subgroup of $\Gamma(2)$. Then Λ is free of infinite rank.

PROBLEM 2.1 (A. Kontorovich). Which integers are traces of elements of Λ? Is it true that the local obstruction is the only obstruction?

Here, $z \in \mathbb{Z}$ is *locally* a trace of an element of Λ provided that for each natural number n, $z \pmod n$ is the trace of an element of Λ (also taken mod n). Note that B. Ogrodnik precisely identified all the local obstructions in this problem [**53**], and studied extensive numerics and other related considerations for this problem. We do not currently know that a positive proportion of numbers arise as traces! For progress on related "local-global"-type problems, see [**7, 8, 41**]. There has also been recent progress on traces in very thin (having critical exponent anything above 1/2) subgroups of $SL(2,\mathbb{Z})$, assuming said subgroups contain parabolic elements (which the above Λ does not): see [**42**].

3. $SL(3,\mathbb{Z})$-related problems

This is a series of general questions about structure of subgroups of $SL(3,\mathbb{Z})$.

3.1. Intrinsic properties of thin subgroups of $SL(3,\mathbb{Z})$.

PROBLEM 3.1. What are finitely generated thin subgroups of $SL(3,\mathbb{Z})$ as abstract groups?

Note that all currently known thin subgroups of $SL(3,\mathbb{Z})$ are either virtually free or virtually surface groups. Examples of free subgroups are given by Tits' ping-pong argument; see [**65**]. Examples of thin surface subgroups of $SL(3,\mathbb{Z})$ are constructed in [**44**].

PROBLEM 3.2 (M. Kapovich). Give an example of a finitely generated thin subgroup of $SL(3,\mathbb{Z})$ which is neither virtually free nor is virtually a surface group. For instance, does the free product of two surface groups embed? Does the free product $\mathbb{Z}^2 \star \mathbb{Z}$ embed?

Note that $SL(4,\mathbb{Z})$ contains subgroups isomorphic to $\mathbb{Z}^2 \star \mathbb{Z}$ and free products of surface groups.

The existence of subgroups of $SL(3,\mathbb{Z})$ isomorphic to $\mathbb{Z}^2 \star \mathbb{Z}$ was claimed by G. Soifer in [**63**], but the proof is known to be wrong. In the subsequent paper [**64**], G. Soifer constructs subgroups in $SL(3,\mathbb{Q})$ isomorphic to $\mathbb{Z}^2 \star \mathbb{Z}$. However, the construction depends on the existence of *singular* diagonalizable elements in $SL(3,\mathbb{Q})\setminus\{1\}$ and such elements do not exist in $SL(3,\mathbb{Z})$. In fact, Soifer's construction requires the existence of *supersingular* diagonalizable subgroups in $SL(3,\mathbb{Q})$ (see section 1).

In view of Soifer's construction, it makes sense to ask a slightly more general question:

PROBLEM 3.3 (M. Kapovich). Does there exist a discrete subgroup $\Gamma < SL(3,\mathbb{R})$ isomorphic to $\mathbb{Z}^2 \star \mathbb{Z}$ and containing only regular diagonalizable elements?

Similarly:

PROBLEM 3.4 (K. Tsouvalas). Does there exist a discrete subgroup $\Gamma < SL(3,\mathbb{R})$ isomorphic to $\Gamma_0 \star \mathbb{Z}$, where Γ_0 is a surface group?

Note that it is impossible to find an Anosov subgroup $\Gamma < SL(3,\mathbb{R})$ isomorphic to $\Gamma_0 \star \mathbb{Z}$ with this property, since every Anosov subgroup of $SL(3,\mathbb{R})$ is either virtually free or a virtually surface group, [**16**].

PROBLEM 3.5. Is $SL(3,\mathbb{Z})$ coherent?

This open problem goes back to Serre (1974), [**62**]. It is known that virtually free groups and virtually surface groups are coherent. More generally, fundamental groups of 3-dimensional manifolds are coherent. In particular, discrete subgroups of $PSL(2,\mathbb{C})$ are coherent. More examples of coherent groups come from combinatorial group theory; see the survey [**70**] by Wise. On the other hand, the groups $SL(n,\mathbb{Z})$, $n \geq 4$, are known to be noncoherent since they contain a copy of $SL(2,\mathbb{Z}) \times SL(2,\mathbb{Z})$, hence, of $F_2 \times F_2$, and the latter is known to be noncoherent; see e.g. [**50**]. (Here F_2 is the rank 2 free group.)

3.2. Extrinsic properties of thin subgroups of $SL(3,\mathbb{Z})$. In their pioneering paper [**47**], Margulis and Soifer proved that every finitely generated matrix group is either virtually polycyclic or contains a strictly maximal subgroup. However, very little is known about the algebraic structure of such subgroups.

PROBLEM 3.6. Is there a virtually free strictly maximal subgroup in $SL(n,\mathbb{Z})$, $n \geq 3$?

Note that the proof of existence of strictly maximal subgroups in the work of Margulis and Soifer starts with construction of a profinitely dense free subgroup. But the next step of the construction is to extend such a subgroup to a maximal subgroup and it is totally unclear what happens to the algebraic structure of the subgroup in the process.

The next problem is due to G. Prasad and J. Tits:

PROBLEM 3.7. Is every strictly maximal subgroup of $SL(3,\mathbb{Z})$ virtually free?

According to Margulis and Soifer [**48**], Prasad and Tits asked this question for $SL(n,\mathbb{Z})$, $n \geq 3$. It was proven by Margulis and Soifer [**48**] that the answer is negative for $n \geq 4$. The remaining open case is for $n = 3$.

The following open problem also goes back to the work of Margulis and Soifer [**47**], where they proved the existence of strictly maximal subgroups in finitely generated non-polycyclic matrix groups:

PROBLEM 3.8. Are there finitely generated strictly maximal subgroups of $SL(3,\mathbb{Z})$? The same question for $SL(n,\mathbb{Z})$, $n \geq 4$. (The expected answer is negative.)

PROBLEM 3.9 (J.-P. Serre). Is there a profinitely dense non-virtually free subgroup in $SL(3,\mathbb{Z})$?

Note that the key ingredient in proof of existence of strictly maximal subgroups in $SL(n,\mathbb{Z})$, $n \geq 4$, given in [**48**], is the existence of profinitely dense subgroups containing $\mathbb{Z}^2$.

PROBLEM 3.10. Is it true that Anosov subgroups of $SL(3,\mathbb{Z})$ are never maximal?

REMARK 3.11. The only known results about nonexistence of strictly maximal finitely generated subgroups are in rank 1:

1. It is an easy consequence of the ping-pong argument that if M is a maximal geometrically finite subgroup of a lattice in a rank 1 Lie group Γ, then M has finite index in Γ.

2. A much harder theorem is that every maximal finitely generated subgroup in a lattice $\Gamma < O(3,1)$ necessarily has finite index in Γ; see [**26**]. (This is an application of deep structural results about finitely generated discrete subgroups of $O(3,1)$.)

REMARK 3.12. It is known that every Anosov surface subgroup Γ of $SL(3,\mathbb{R})$ is *virtually* a maximal Anosov subgroup, i.e. if $\Lambda < SL(3,\mathbb{R})$ is any Anosov subgroup containing Γ, then $|\Gamma : \Lambda| < \infty$. At the same time, free Anosov subgroups of a semisimple Lie group G are never virtually maximal as Anosov subgroups, cf. [**20**].

PROBLEM 3.13. Does $SL(3,\mathbb{Z})$ have the Howson property?

REMARK 3.14.

1. The answer to the previous question is negative for $SL(n,\mathbb{Z})$, $n \geq 4$, and positive for $SL(2,\mathbb{Z})$; see Section 1.

2. The Howson property is unclear even for intersections of Anosov subgroups of $SL(3,\mathbb{Z})$; cf. the discussion in Section 1.

PROBLEM 3.15. Is there a lattice $\Gamma < SL(3,\mathbb{R})$ containing a singular diagonal element?

Note that such a lattice will also necessarily contain a product subgroup $F_2 \times \mathbb{Z}$. Then this lattice will not have the Howson property with respect to Anosov subgroups; see Example 1.1.

PROBLEM 3.16. Give an example of a finitely generated thin subgroup of $SL(3,\mathbb{Z})$ which is not (relatively) Anosov.

Note that all currently known constructions of finitely-generated thin subgroups of $SL(3,\mathbb{Z})$ are relatively Anosov. It is known that all finitely generated discrete subgroups of $SL(2,\mathbb{R})$ are geometrically finite, hence, relatively Anosov.

The next problem is motivated by the Howson property: It is possible that it is easier to prove this property by restricting to the class of Anosov subgroups:

PROBLEM 3.17 (M. Kapovich). Suppose that Γ_1, Γ_2 are Anosov subgroups of $SL(3,\mathbb{Z})$. Is $\Gamma_1 \cap \Gamma_2$ finitely generated?

4. Problems on higher rank lattices

4.1. Profinite density.

PROBLEM 4.1 (G. Soifer). Does there exist a thin profinitely dense subgroup of $SL(n,\mathbb{Z})$, $n \geq 3$, generated by two elements?

Note that Aka, Gelander and Soifer [**1**] proved that there exists a uniform constant k such that for every n, $SL(n,\mathbb{Z})$ contains a k-generated thin profinitely dense subgroup.

4.2. Commutator map problems.

PROBLEM 4.2 (A. Shalev). Is it true that for $n \geq 3$ the commutator map of $SL(n,\mathbb{Z})$ is surjective?

Note that $SL(n,\mathbb{Z})$ is a perfect group (equal to its own commutator subgroup), which implies that every element is a product of commutators. Not every perfect group has surjective commutator map. One measure of failure of surjectivity of the commutator map in a group Γ is given by the *commutator length* and *stable commutator length*:

Given $\gamma \in [\Gamma,\Gamma]$, let $\ell(\gamma)$ denote the least number k such that γ is the product of k commutators in Γ. The number $\ell(\gamma)$ is called the *commutator length* of γ. This quantity has an asymptotic counterpart, the *stable commutator length*:

$$\ell_\infty(\gamma) = \lim_{n\to\infty} \frac{\ell(\gamma^n)}{n}.$$

There are perfect groups which have elements of positive stable commutator length: For instance, each hyperbolic van Dyck group with the presentation

$$\langle a,b,c \mid a^p = b^q = c^r = abc = 1\rangle, \quad p^{-1}+q^{-1}+r^{-1} < 1$$

is perfect whenever the numbers p, q, r are pairwise coprime. However, such a group (as any nonelementary hyperbolic group) contains elements of positive stable commutator length since it admits unbounded quasimorphisms, [**4**, **22**]. At the same time, if Γ is a lattice in a simple Lie group of rank ≥ 2, then $\ell_\infty(\Gamma) = \{0\}$; see [**15**].

REMARK 4.3. For a group Γ, a map $f : \Gamma \to \mathbb{R}$ is said to be a *quasimorphism* if there is a constant C such that for all $\alpha, \beta \in \Gamma$,

$$|f(\alpha\beta) - f(\alpha) - f(\beta)| \leq C.$$

In other words, quasimorphisms are *approximate* additive characters of a group. Trivial examples of quasimorphisms are given by bounded maps $f : \Gamma \to \mathbb{R}$. Quasimorphisms form a real vector space. The quotient, denoted $QM(\Gamma)$, of this space by the subspace of bounded quasimorphisms detects "richness" of the space of quasimorphisms of Γ. There are many groups which do not admit nontrivial additive characters, but do admit unbounded quasimorphisms. For instance, for every *nonelementary hyperbolic group* Γ, the space $QM(\Gamma)$ is infinite-dimensional, [**22**].

Note also that if $\Gamma = SL(2, \mathcal{O}_K)$, where $\mathcal{O}_K$ is the ring of integers of a quadratic field with infinitely many units, then Γ contains elements which are commutators *locally but not globally* [**24**]. Here, an element γ is *locally a commutator* if its image in every congruence-quotient of Γ is a commutator. An element of Γ is a commutator *globally* if it equals the commutator $[\alpha, \beta]$ for some $\alpha, \beta \in \Gamma$. Clearly, every global commutator is also a local commutator.

4.3. Characterization of higher rank lattices.

PROBLEM 4.4 (M. Kapovich). What algebraic properties distinguish higher rank (irreducible uniform) lattices among abstract groups?

One such characterization was given by Lubotzky and Venkataramana [**46**], in terms of profinite completions. There are some indirect signs that other algebraic characterizations of lattices are also possible:

(1) Higher rank lattices are quasi-isometrically rigid (Kleiner and Leeb [**40**], Eskin [**23**]).
(2) Higher rank lattices are rigid in the sense of the 1st order logic (Avni, Lubotzky, Mieri [**2**]).
(3) Appearance of Serre relators in profinite completions (Prasad, Rapinchuk [**58**]).

In the case of groups Γ of integer points of split semisimple algebraic groups over $\mathbb{Z}$, a defining feature is the *Serre relators*. However, Serre relators are for unipotent elements, which do not exist in uniform lattices. Uniform higher rank lattices satisfy *approximate* Serre relators.

PROBLEM 4.5. Do these determine whether a discrete linear group is a higher rank lattice?

An alternative approach to a characterization of lattices is via the *Prasad–Raghunathan rank*:

DEFINITION 4.1 (Prasad–Raghunathan rank). Let Γ be a group. Let A_i denote the subset of Γ that consists of those elements whose centralizer contains a free abelian group of rank at most i as a subgroup of finite index. Thus, $A_0 \subset A_1 \subset \ldots$.

The *Prasad–Raghunathan rank*, PRrank(Γ), of Γ is the minimal number i such that $\Gamma = \gamma_1 A_i \cup \cdots \cup \gamma_m A_i$ for some $\gamma_1, \ldots, \gamma_m \in \Gamma$.

For instance, if Γ is a lattice in a semisimple Lie group of rank n, then PRrank(Γ) $= n$. If M is a compact Riemannian manifold of nonpositive curvature with $\Gamma = \pi_1(M)$, then PRrank(Γ) equals the geometric rank of M, i.e. the largest n such that every geodesic in M is contained in an immersed n-dimensional flat. We refer to **[56]** and **[3]** for details.

PROBLEM 4.6. Are there discrete linear groups Γ which are not virtually nontrivial direct products and are not lattices, satisfying PRrank(Γ) ≥ 2?

PROBLEM 4.7 (G. Prasad). Does there exist a discrete Zariski dense subgroup $\Gamma < G$ (with G a simple real algebraic group) such that Γ is not a lattice but PRrank(Γ) $= rank_{\mathbb{R}}(G)$?

Another group-theoretic property closely related to lattices is the *bounded generation property*:

DEFINITION 4.2 (BGP, Bounded Generation Property). A group Γ is said to have BGP if there exist elements $\gamma_1, \ldots, \gamma_k$ such that every $\gamma \in \Gamma$ can be written as a product

$$\gamma = \gamma_1^{n_1} \gamma_2^{n_2} \cdots \gamma_k^{n_k}$$

for some $n_1, \ldots, n_k \in \mathbb{Z}$. (Note that a power of each γ_i appears only once.)

Many classes of higher rank nonuniform lattices satisfy the BGP; see the references in **[17]**. Nonlinear groups that satisfy the BGP were constructed by A. Muranov **[52]**. On the other hand, it was recently proven in **[17]** that uniform lattices in semisimple Lie groups *never* satisfy the BGP. More generally, they prove that a subgroup of $SL(n, \mathbb{C})$ boundedly generated by semisimple elements has to be virtually solvable.

PROBLEM 4.8 (M. Kapovich). Suppose that Γ is an abstract (infinite) $\mathbb{R}$-linear group satisfying the BGP. Is it isomorphic to a lattice in a Lie group?

4.4. Why are higher rank lattices super-rigid? One way to say that an abstract group Γ is *super-rigid* is to require that for every field F and $n \in \mathbb{N}$, there are only finitely many conjugacy classes of representations $\Gamma \to GL(n, F)$. Of course, some groups do not admit any nontrivial linear representations, so it makes sense to restrict the discussion to finitely generated linear groups Γ.

Loosely speaking, such a group is (super) rigid if it satisfies some peculiar relators. There are many proofs of rigidity and super-rigidity of (higher rank irreducible) lattices, but none of these proofs (in the setting of uniform lattices) use relators satisfied by lattices, likely because such relators are simply unknown (see previous section). In contrast, there are known proofs of super-rigidity of some classes of higher rank non-uniform lattices (see **[60]** and references therein) which use explicit relators.

PROBLEM 4.9 (M. Kapovich). What are group-theoretic reasons that make higher rank uniform lattices (super)-rigid? Are the approximate Serre relators responsible for this? Or high Prasad-Raghunathan rank?

One known result in this direction is that the BGP implies super-rigidity **[55]**. Another group-theoretic property implying super-rigidity is given by Lubotzky in **[45]**.

5. Algorithmic problems

PROBLEM 5.1 (M. Kapovich). For which classes of algebraic semisimple Lie groups G is the discreteness problem decidable for Zariski dense finitely generated subgroups?

Here, decidability of discreteness is understood in the sense of BSS formalism of computations over the real numbers, as it is discussed for instance in [**25**] and [**33**]. The input for a possible BSS algorithm consists of a finite tuple of elements of G which generate a Zariski dense subgroup. The algorithm is supposed to determine if these elements generate a discrete subgroup. The Zariski density assumption is imposed to eliminate "trivial" counter-examples, which show that discreteness is undecidable already in the case of cyclic subgroups of S^1. It is known that discreteness is decidable for finitely generated subgroups of $G = PSL(2,\mathbb{R})$ (see for instance, Gilman's paper [**25**] and references therein) and is undecidable for subgroups of $PSL(2,\mathbb{C})$ (see [**33**]). The simplest case where the answer is unclear is, as usual, $G = SL(3,\mathbb{R})$.

PROBLEM 5.2 (A. Detinko). Is freeness decidable for finitely generated subgroups of arithmetic groups?

There is a practical algorithm testing whether a finitely generated linear group over an arbitrary (infinite) field contains a free non-abelian subgroup ([**19**, Section 6.2]).

Note that freeness is undecidable for subsemigroups in linear groups; see [**39**]. Freeness is decidable for subgroups of $SL(2,\mathbb{Z})$ and, more generally, for discrete subgroups of $SL(2,\mathbb{R})$. It is also decidable for some special classes of subgroups of arithmetic groups:

(a) Anosov subgroups.

(b) Subgroups which admit finitely-sided Dirichlet domains in associated symmetric spaces.

Freeness is likely to be, at least effectively, undecidable. The reason is the existence of *badly distorted* finitely generated free subgroups of $SL(n,\mathbb{Z})$ for large n: these are free subgroups whose distortion function is comparable to the k-th Ackermann function (for any k); see [**13, 21**] for the description of embeddings of such free groups in free-by-cyclic groups, and [**29, 69**] for embeddings into $SL(n,\mathbb{Z})$.

PROBLEM 5.3 (A. Detinko). Is arithmeticity decidable? More precisely, is there an algorithm that decides if a finitely generated Zariski dense subgroup Λ (given by its set of generators) of an *irreducible* arithmetic group Γ (say, $SL(n,\mathbb{Z})$, $n \geq 3$) has finite index in Γ (cf. [**19**, Section 5.3])?

Note that this problem is semidecidable: There is an algorithm which will terminate if $\Lambda < \Gamma$ has finite index. The problem is known to be decidable for subgroups of $SL(2,\mathbb{Z})$ and undecidable for subgroups of $SL(2,\mathbb{Z}) \times SL(2,\mathbb{Z})$.

PROBLEM 5.4 (M. Kapovich). Is the membership problem for finitely generated subgroups of $SL(3,\mathbb{Z})$ decidable?

Note that the membership problem for a finitely generated subgroup H of a finitely generated group G is decidable if and only if the *distortion function* of H in G is recursive. All *known* finitely generated subgroups of $SL(3,\mathbb{Z})$ have at most exponential distortion, hence, have decidable membership problem.

In contrast, the membership problem is undecidable for finitely generated subgroups of $SL(4,\mathbb{Z})$. The reason is that this group contains $SL(2,\mathbb{Z}) \times SL(2,\mathbb{Z})$, which, in turn, contains a direct product of two free groups of large ranks. The latter contains finitely generated normal subgroups with undecidable membership problem (*Mihailova subgroups*, [**50**]). However, in this case, the ambient lattice is reducible.

PROBLEM 5.5. Are there irreducible arithmetic groups Γ such that for Zariski dense subgroups $\Lambda < \Gamma$ the membership problem is undecidable?

Very likely, such arithmetic subgroups Γ can be found in $SO(p,q)$ for suitable p,q. The existence of Λ is an application of the Rips construction of small cancellation groups with non-recursively distorted normal subgroups [**61**], combined with the Cubulation Theorem of Dani Wise [**68**] and the embedability of cubulated groups in RACGs (Right-Angled Coxeter groups) [**69**], which, in turn, admit Zariski dense representations in $\Gamma := O(p,q) \cap GL(p+q,\mathbb{Z})$ [**5**].

Recall that the membership problem is decidable for quasi-isometrically embedded subgroups, such as Anosov subgroups and finite-index subgroups in lattices.

PROBLEM 5.6. Suppose that Γ is an irreducible lattice in a higher rank semisimple Lie group. Is it decidable that $\gamma \in \Gamma$ is a commutator?

Note that this question is a special case of decidability of equations in Γ. In the last 20 or so years there was a great deal of progress in understanding equations in (relatively) hyperbolic groups (which includes lattices in rank 1 Lie groups). In contrast, decidability of equations in higher rank lattices is very poorly understood.

Here is a similar number-theoretic problem:

PROBLEM 5.7. Is every integer $n \in \mathbb{Z}$ a sum of three cubes, where n is not 4 nor 5 modulo 9? Is it even decidable if the given integer is a sum of three cubes?

References

[1] Menny Aka, Tsachik Gelander, and Gregory A. Soĭfer, *Homogeneous number of free generators*, J. Group Theory **17** (2014), no. 4, 525–539, DOI 10.1515/jgt-2014-0001. MR3228932

[2] Nir Avni, Alexander Lubotzky, and Chen Meiri, *First order rigidity of non-uniform higher rank arithmetic groups*, Invent. Math. **217** (2019), no. 1, 219–240, DOI 10.1007/s00222-019-00866-5. MR3958794

[3] Werner Ballmann and Patrick Eberlein, *Fundamental groups of manifolds of nonpositive curvature*, J. Differential Geom. **25** (1987), no. 1, 1–22. MR873453

[4] Christophe Bavard, *Longueur stable des commutateurs* (French), Enseign. Math. (2) **37** (1991), no. 1-2, 109–150. MR1115747

[5] Yves Benoist and Pierre de la Harpe, *Adhérence de Zariski des groupes de Coxeter* (French, with English summary), Compos. Math. **140** (2004), no. 5, 1357–1366, DOI 10.1112/S0010437X04000338. MR2081159

[6] Yves Benoist and Hee Oh, *Discrete subgroups of* $\mathrm{SL}_3(\mathbb{R})$ *generated by triangular matrices*, Int. Math. Res. Not. IMRN **4** (2010), 619–632, DOI 10.1093/imrn/rnp149. MR2595007

[7] Jean Bourgain and Alex Kontorovich, *On the local-global conjecture for integral Apollonian gaskets*, Invent. Math. **196** (2014), no. 3, 589–650, DOI 10.1007/s00222-013-0475-y. With an appendix by Péter P. Varjú. MR3211042

[8] Jean Bourgain and Alex Kontorovich, *On Zaremba's conjecture*, Ann. of Math. (2) **180** (2014), no. 1, 137–196, DOI 10.4007/annals.2014.180.1.3. MR3194813

[9] Jean Bourgain, Alex Gamburd, and Peter Sarnak, *Affine linear sieve, expanders, and sum-product*, Invent. Math. **179** (2010), no. 3, 559–644, DOI 10.1007/s00222-009-0225-3. MR2587341

[10] Jean Bourgain and Alex Kontorovich, *On representations of integers in thin subgroups of* $SL_2(\mathbb{Z})$, Geom. Funct. Anal. **20** (2010), no. 5, 1144–1174, DOI 10.1007/s00039-010-0093-4. MR2746949

[11] Jean Bourgain and Péter P. Varjú, *Expansion in* $SL_d(\mathbf{Z}/q\mathbf{Z})$, *q arbitrary*, Invent. Math. **188** (2012), no. 1, 151–173, DOI 10.1007/s00222-011-0345-4. MR2897695

[12] B. H. Bowditch, *Geometrical finiteness with variable negative curvature*, Duke Math. J. **77** (1995), no. 1, 229–274, DOI 10.1215/S0012-7094-95-07709-6. MR1317633

[13] Noel Brady, Will Dison, and Timothy Riley, *Hyperbolic hydra*, Groups Geom. Dyn. **7** (2013), no. 4, 961–976, DOI 10.4171/GGD/212. MR3134032

[14] E. Breuillard and T. Gelander, *Uniform independence in linear groups*, Invent. Math. **173** (2008), no. 2, 225–263, DOI 10.1007/s00222-007-0101-y. MR2415307

[15] M. Burger and N. Monod, *Bounded cohomology of lattices in higher rank Lie groups*, J. Eur. Math. Soc. (JEMS) **1** (1999), no. 2, 199–235, DOI 10.1007/s100970050007. MR1694584

[16] Richard Canary and Konstantinos Tsouvalas, *Topological restrictions on Anosov representations*, J. Topol. **13** (2020), no. 4, 1497–1520, DOI 10.1112/topo.12166. MR4186136

[17] Pietro Corvaja, Andrei S. Rapinchuk, Jinbo Ren, and Umberto M. Zannier, *Non-virtually abelian anisotropic linear groups are not boundedly generated*, Invent. Math. **227** (2022), no. 1, 1–26, DOI 10.1007/s00222-021-01064-y. MR4359474

[18] Jeffrey Danciger, François Guéritaud, and Fanny Kassel, *Convex cocompactness in pseudo-Riemannian hyperbolic spaces*, Geom. Dedicata **192** (2018), 87–126, DOI 10.1007/s10711-017-0294-1. MR3749424

[19] A. S. Detinko and D. L. Flannery, *Linear groups and computation*, Expo. Math. **37** (2019), no. 4, 454–484, DOI 10.1016/j.exmath.2018.07.002. MR4040183

[20] Subhadip Dey, Michael Kapovich, and Bernhard Leeb, *A combination theorem for Anosov subgroups*, Math. Z. **293** (2019), no. 1-2, 551–578, DOI 10.1007/s00209-018-2208-9. MR4002289

[21] Will Dison and Timothy R. Riley, *Hydra groups*, Comment. Math. Helv. **88** (2013), no. 3, 507–540, DOI 10.4171/CMH/294. MR3093501

[22] David B. A. Epstein and Koji Fujiwara, *The second bounded cohomology of word-hyperbolic groups*, Topology **36** (1997), no. 6, 1275–1289, DOI 10.1016/S0040-9383(96)00046-8. MR1452851

[23] Alex Eskin, *Quasi-isometric rigidity of nonuniform lattices in higher rank symmetric spaces*, J. Amer. Math. Soc. **11** (1998), no. 2, 321–361, DOI 10.1090/S0894-0347-98-00256-2. MR1475886

[24] Amit Ghosh, Chen Meiri, and Peter Sarnak, *Commutators in* SL_2 *and Markoff surfaces I*, New Zealand J. Math. **52** (2021), 773–819, DOI 10.53733/198. MR4387994

[25] Jane Gilman, *Algorithms, complexity and discreteness criteria in* $PSL(2, \mathbf{C})$, J. Anal. Math. **73** (1997), 91–114, DOI 10.1007/BF02788139. MR1616469

[26] Yair Glasner, Juan Souto, and Peter Storm, *Finitely generated subgroups of lattices in* $PSL_2\,\mathbb{C}$, Proc. Amer. Math. Soc. **138** (2010), no. 8, 2667–2676, DOI 10.1090/S0002-9939-10-10310-4. MR2644883

[27] Leon Greenberg, *Discrete groups of motions*, Canadian J. Math. **12** (1960), 415–426, DOI 10.4153/CJM-1960-036-8. MR115130

[28] Olivier Guichard and Anna Wienhard, *Anosov representations: domains of discontinuity and applications*, Invent. Math. **190** (2012), no. 2, 357–438, DOI 10.1007/s00222-012-0382-7. MR2981818

[29] Mark F. Hagen and Daniel T. Wise, *Cubulating hyperbolic free-by-cyclic groups: the general case*, Geom. Funct. Anal. **25** (2015), no. 1, 134–179, DOI 10.1007/s00039-015-0314-y. MR3320891

[30] John Hempel, *The finitely generated intersection property for Kleinian groups*, Knot theory and manifolds (Vancouver, B.C., 1983), Lecture Notes in Math., vol. 1144, Springer, Berlin, 1985, pp. 18–24, DOI 10.1007/BFb0075010. MR823280

[31] A. G. Howson, *On the intersection of finitely generated free groups*, J. London Math. Soc. **29** (1954), 428–434, DOI 10.1112/jlms/s1-29.4.428. MR65557

[32] Michael Kapovich, *Kleinian groups in higher dimensions*, Geometry and dynamics of groups and spaces, Progr. Math., vol. 265, Birkhäuser, Basel, 2008, pp. 487–564, DOI 10.1007/978-3-7643-8608-5_13. MR2402415

[33] Michael Kapovich, *Discreteness is undecidable*, Internat. J. Algebra Comput. **26** (2016), no. 3, 467–472, DOI 10.1142/S0218196716500193. MR3506344

[34] Michael Kapovich, *A survey of complex hyperbolic Kleinian groups*, In the tradition of Thurston II. Geometry and groups, Springer, Cham, 2022, pp. 7–51, DOI 10.1007/978-3-030-97560-9_2. MR4472047

[35] M. Kapovich, *Geometric algorithms for discreteness and faithfulness*, Computational Aspects of Discrete Subgroups of Lie Groups (Detinko et al., Eds.), Contemporary Mathematics, vol. 783, Amer. Math. Soc., Providence, RI, 2023.

[36] M. Kapovich, B. Leeb, *Relativizing characterizations of Anosov subgroups, I.* Preprint, `arXiv:1807.00160`, 2018.

[37] M. Kapovich, B. Leeb, and J. Porti, *Morse actions of discrete groups on symmetric spaces*, Preprint, `arXiv:1403.7671`, 2014.

[38] Michael Kapovich, Bernhard Leeb, and Joan Porti, *Some recent results on Anosov representations*, Transform. Groups **21** (2016), no. 4, 1105–1121, DOI 10.1007/s00031-016-9393-6. MR3569569

[39] David A. Klarner, Jean-Camille Birget, and Wade Satterfield, *On the undecidability of the freeness of integer matrix semigroups*, Internat. J. Algebra Comput. **1** (1991), no. 2, 223–226, DOI 10.1142/S0218196791000146. MR1128014

[40] Bruce Kleiner and Bernhard Leeb, *Rigidity of quasi-isometries for symmetric spaces and Euclidean buildings*, Inst. Hautes Études Sci. Publ. Math. **86** (1997), 115–197. MR1608566

[41] Alex Kontorovich, *From Apollonius to Zaremba: local-global phenomena in thin orbits*, Bull. Amer. Math. Soc. (N.S.) **50** (2013), no. 2, 187–228, DOI 10.1090/S0273-0979-2013-01402-2. MR3020826

[42] A. Kontorovich and X. Zhang, *On length sets of subarithmetic hyperbolic manifolds*, Preprint, `arXiv:2201.10955`, 2022.

[43] D. D. Long and A. W. Reid, *Thin surface subgroups in cocompact lattices in* $\mathrm{SL}(3,\mathbf{R})$, Illinois J. Math. **60** (2016), no. 1, 39–53. MR3665171

[44] Darren D. Long, Alan W. Reid, and Morwen Thistlethwaite, *Zariski dense surface subgroups in* $\mathrm{SL}(3,\mathbf{Z})$, Geom. Topol. **15** (2011), no. 1, 1–9, DOI 10.2140/gt.2011.15.1. MR2764111

[45] Alexander Lubotzky, *Dimension function for discrete groups*, Proceedings of groups—St. Andrews 1985, London Math. Soc. Lecture Note Ser., vol. 121, Cambridge Univ. Press, Cambridge, 1986, pp. 254–262. MR896520

[46] A. Lubotzky and T. N. Venkataramana, *A group theoretical characterisation of S-arithmetic groups in higher rank semi-simple groups*, Geom. Dedicata **90** (2002), 1–28, DOI 10.1023/A:1014985620796. MR1898148

[47] G. A. Margulis and G. A. Soĭfer, *A criterion for the existence of maximal subgroups of infinite index in a finitely generated linear group* (Russian), Dokl. Akad. Nauk SSSR **234** (1977), no. 6, 1261–1264. MR0466412

[48] G. A. Margulis and G. A. Soĭfer, *Nonfree maximal subgroups of infinite index of the group* $\mathrm{SL}_n(\mathbf{Z})$ (Russian), Uspekhi Mat. Nauk **34** (1979), no. 4 (208), 203–204. MR548436

[49] G. A. Margulis and G. A. Soĭfer, *Maximal subgroups of infinite index in finitely generated linear groups*, J. Algebra **69** (1981), no. 1, 1–23, DOI 10.1016/0021-8693(81)90123-X. MR613853

[50] K. A. Mihaĭlova, *The occurrence problem for free products of groups* (Russian), Mat. Sb. (N.S.) **75 (117)** (1968), 199–210. MR0222179

[51] D. I. Moldavanskiĭ, *The intersection of finitely generated subgroups* (Russian), Sibirsk. Mat. Ž. **9** (1968), 1422–1426. MR0237619

[52] Alexey Muranov, *Diagrams with selection and method for constructing boundedly generated and boundedly simple groups*, Comm. Algebra **33** (2005), no. 4, 1217–1258, DOI 10.1081/AGB-200053951. MR2136699

[53] Brooke Logan Ogrodnik, *On the Local-Global Conjecture for Commutator Traces*, ProQuest LLC, Ann Arbor, MI, 2021. Thesis (Ph.D.)–Rutgers The State University of New Jersey, School of Graduate Studies. MR4314570

[54] Hee Oh, *Discrete subgroups generated by lattices in opposite horospherical subgroups*, J. Algebra **203** (1998), no. 2, 621–676, DOI 10.1006/jabr.1997.7188. MR1622748

[55] Vladimir Platonov and Andrei Rapinchuk, *Algebraic groups and number theory*, Pure and Applied Mathematics, vol. 139, Academic Press, Inc., Boston, MA, 1994. Translated from the 1991 Russian original by Rachel Rowen. MR1278263

[56] Gopal Prasad and M. S. Raghunathan, *Cartan subgroups and lattices in semi-simple groups*, Ann. of Math. (2) **96** (1972), 296–317, DOI 10.2307/1970790. MR302822

[57] Gopal Prasad and Andrei S. Rapinchuk, *Weakly commensurable arithmetic groups and isospectral locally symmetric spaces*, Publ. Math. Inst. Hautes Études Sci. **109** (2009), 113–184, DOI 10.1007/s10240-009-0019-6. MR2511587

[58] Gopal Prasad and Andrei S. Rapinchuk, *Generic elements in Zariski-dense subgroups and isospectral locally symmetric spaces*, Thin groups and superstrong approximation, Math. Sci. Res. Inst. Publ., vol. 61, Cambridge Univ. Press, Cambridge, 2014, pp. 211–252. MR3220892

[59] Gopal Prasad and Andrei S. Rapinchuk, *On the congruence kernel for simple algebraic groups*, Tr. Mat. Inst. Steklova **292** (2016), no. Algebra, Geometriya i Teoriya Chisel, 224–254, DOI 10.1134/S0371968516010143; English transl., Proc. Steklov Inst. Math. **292** (2016), no. 1, 216–246. MR3628463

[60] Igor A. Rapinchuk, *On abstract homomorphisms of Chevalley groups over the coordinate rings of affine curves*, Transform. Groups **24** (2019), no. 4, 1241–1259, DOI 10.1007/s00031-018-9494-5. MR4038092

[61] E. Rips, *Subgroups of small cancellation groups*, Bull. London Math. Soc. **14** (1982), no. 1, 45–47, DOI 10.1112/blms/14.1.45. MR642423

[62] J.-P. Serre, *Problems.* In: "Proceedings of the Second International Conference on the Theory of Groups", edited by M. F. Newman, pp. 734–735, Lecture Notes in Mathematics, **372**, Springer-Verlag, Berlin-New York, 1974.

[63] G. A. Soifer, *Free subgroups of linear groups*, Pure Appl. Math. Q. **3** (2007), no. 4, Special Issue: In honor of Grigory Margulis., 987–1003, DOI 10.4310/PAMQ.2007.v3.n4.a6. MR2402593

[64] G. A. Soifer, *Embedding* $\mathbb{Z}^2 * Z$ *as a discrete subgroup into small linear groups*, Internat. J. Algebra Comput. **22** (2012), no. 7, 1250057, 10, DOI 10.1142/S0218196712500579. MR2999365

[65] J. Tits, *Free subgroups in linear groups*, J. Algebra **20** (1972), 250–270, DOI 10.1016/0021-8693(72)90058-0. MR286898

[66] T. N. Venkataramana, *Zariski dense subgroups of arithmetic groups*, J. Algebra **108** (1987), no. 2, 325–339, DOI 10.1016/0021-8693(87)90106-2. MR892908

[67] T. N. Venkataramana, *On systems of generators of arithmetic subgroups of higher rank groups*, Pacific J. Math. **166** (1994), no. 1, 193–212. MR1306038

[68] D. T. Wise, *Cubulating small cancellation groups*, Geom. Funct. Anal. **14** (2004), no. 1, 150–214, DOI 10.1007/s00039-004-0454-y. MR2053602

[69] Daniel T. Wise, *From riches to raags: 3-manifolds, right-angled Artin groups, and cubical geometry*, CBMS Regional Conference Series in Mathematics, vol. 117, Published for the Conference Board of the Mathematical Sciences, Washington, DC; by the American Mathematical Society, Providence, RI, 2012, DOI 10.1090/cbms/117. MR2986461

[70] Daniel T. Wise, *An invitation to coherent groups*, What's next?—the mathematical legacy of William P. Thurston, Ann. of Math. Stud., vol. 205, Princeton Univ. Press, Princeton, NJ, 2020, pp. 326–414, DOI 10.2307/j.ctvthhdvv.16. MR4205645

[71] Andrew Zimmer, *Projective Anosov representations, convex cocompact actions, and rigidity*, J. Differential Geom. **119** (2021), no. 3, 513–586, DOI 10.4310/jdg/1635368438. MR4333029

Department of Mathematics, University of California, Davis, California 95616
Email address: `kapovich@ucdavis.edu`

Department of Computer Science, University of Huddersfield, United Kingdom
Email address: `a.detinko@hud.ac.uk`

Rutgers University, New Brunswick, New Jersey; and National Museum of Mathematics, New York, New York
Email address: `alex.kontorovich@rutgers.edu`

Contemporary Mathematics
Volume **783**, 2023
https://doi.org/10.1090/conm/783/15735

Picard modular groups generated by complex reflections

Alice Mark, Julien Paupert, and David Polletta

ABSTRACT. In this short note we use the presentations found by the various authors to show that the Picard modular groups $\mathrm{PU}(2,1,\mathcal{O}_d)$ with $d = 1, 3, 7$ (respectively the quaternion hyperbolic lattice $\mathrm{PSp}(2,1,\mathcal{H})$ with entries in the Hurwitz integer ring $\mathcal{H}$) are generated by complex (resp. quaternionic) reflections, and that the Picard modular groups $\mathrm{PU}(2,1,\mathcal{O}_d)$ with $d = 2, 11$ have an index 4 subgroup generated by complex reflections.

1. Introduction

Hyperbolic reflection groups are an important class of groups in the realm of discrete subgroups and lattices in Lie groups, and more generally of discrete groups in geometry and topology. Such groups are accessible to a direct geometric description and understanding which are not always clear for groups defined algebraically or arithmetically. While these reflection groups are relatively well understood in the constant curvature setting (they are then *Coxeter groups* in Euclidean, spherical or real hyperbolic n-space), very little is known about their complex and quaternion hyperbolic counterparts.

In the constant curvature setting, reflections are involutions whose fixed-point set is a totally geodesic submanifold of codimension 1. Such submanifolds do not exist in complex or quaternion hyperbolic space of dimension at least 2 (see [**CG**]). Potential substitutes among isometries of complex hyperbolic n-space $\mathrm{H}^n_{\mathbb{C}}$ are *complex reflections*, which are holomorphic isometries fixing pointwise a totally geodesic complex hypersurface (a copy of $\mathrm{H}^{n-1}_{\mathbb{C}} \subset \mathrm{H}^n_{\mathbb{C}}$), and *real reflections*, which are antiholomorphic involutions fixing pointwise a Lagrangian subspace (a copy of $\mathrm{H}^n_{\mathbb{R}} \subset \mathrm{H}^n_{\mathbb{C}}$). The situation in quaternion hyperbolic space is similar, with all totally geodesic subspaces being copies of lower-dimensional real, complex or quaternion hyperbolic spaces.

Major open questions about the existence of lattices generated by real or complex reflections in $\mathrm{PU}(n,1) \simeq \mathrm{Isom}^0(\mathrm{H}^n_{\mathbb{C}})$ include the following. Do there exist lattices generated by (real or complex) reflections in $\mathrm{PU}(n,1)$ for all $n \geqslant 2$? For fixed $n \geqslant 2$, are there infinitely many (non-commensurable) lattices generated by (real or complex) reflections? In real hyperbolic space the answer to the first question is no by classical results of Vinberg (though sharp bounds on possible dimensions are far from known), whereas 2 and 3 are the only dimensions where infinitely many

2020 *Mathematics Subject Classification.* Primary 20F55.

Second author partially supported by National Science Foundation Grant DMS-1708463.

non-commensurable lattices generated by reflections are known to exist, by classical results of Poincaré and Andreev respectively.

In the complex hyperbolic case, Deligne–Mostow ([**DM**]) and Mostow ([**Mos**]) produced lattices generated by complex reflections in $\mathrm{PU}(n,1)$ for all $n \leqslant 9$ (finitely many for each n). Allcock found a further example in $\mathrm{PU}(13,1)$ in [**Al2**], related to the Leech lattice. In the quaternion case Allcock also produced in [**Al1**] and [**Al2**] lattices generated by quaternionic reflections in $\mathrm{PSp}(n,1) \simeq \mathrm{Isom}^0(\mathrm{H}^n_{\mathbb{H}})$ in dimensions $n = 2,3,5,7$, including the Hurwitz lattice in $\mathrm{PSp}(2,1)$ which we consider here (at least up to commensurability).

Stover showed that, among arithmetic lattices in $\mathrm{PU}(n,1)$ ($n \geqslant 2$), only those of *first type* can contain complex reflections (Theorem 1.4 of [**St**], see also Example 9.2 of [**BFMS**]). This class includes the Picard modular groups studied in this note. In particular there exist lattices in $\mathrm{PU}(n,1)$ for all $n \geqslant 2$ which do not contain a single complex reflection (even up to commensurability) - the arithmetic lattices of *second type*, which in dimension 2 contains the fundamental groups of the so-called fake projective planes studied by Klingler ([**K**]) and Prasad–Yeung ([**PY**]) and classified by Cartwright–Steger ([**CS**]). At the other extreme, it turns out that all known non-arithmetic lattices in $\mathrm{PU}(n,1)$ with $n \geqslant 2$ are commensurable to a lattice generated by complex reflections (there are 22 commensurability classes known for $n=2$ and two when $n=3$, see [**DPP**] and [**D**]).

The results in this note contribute to the small list of lattices in $\mathrm{PU}(2,1)$ known to be generated by complex reflections, among Picard modular groups with small discriminant. The Picard modular groups $\mathrm{PU}(2,1,\mathcal{O}_d)$ are the simplest kind of arithmetic lattices in $\mathrm{PU}(2,1)$, analogous to the Bianchi groups $\mathrm{PSL}(2,\mathcal{O}_d)$ in $\mathrm{PSL}(2,\mathbb{C})$. (We denote by $\mathcal{O}_d$ the ring of integers of $\mathbb{Q}[\sqrt{-d}]$, where d is a square-free positive integer). Bianchi proved in the seminal paper [**Bi**] that the Bianchi groups are *reflective*, i.e. generated by reflections up to finite index, for $d \leqslant 19$, $d \neq 14,17$. At the end of the 1980's, Shaiheev extended these results in [**Sh**], using results of Vinberg, proving that only finitely many of the Bianchi groups are reflective, including those with $d \leqslant 21$, $d \neq 14,17$. (The finiteness result now follows from a result of Agol, [**Ag**]). The full classification of reflective Bianchi groups was obtained more recently in [**BeMc**].

The second author and Will proved in [**PWi**] that the Picard modular groups $\mathrm{PU}(2,1,\mathcal{O}_d)$ are generated by real reflections when $d = 1,2,3,7,11$. Our main result is the following:

THEOREM 1.1. *The Picard modular groups* $\mathrm{PU}(2,1,\mathcal{O}_d)$ *with* $d = 1,3,7$ *are generated by complex reflections; when* $d = 2,11$ *they have an index 4 subgroup generated by complex reflections. The Hurwitz modular group* $\mathrm{PSp}(2,1,\mathcal{H})$ *is generated by quaternionic reflections.*

This result was known, at least up to finite index, for the Picard groups with $d = 1,3$ and the Hurwitz group by work of Allcock [**Al1**] (and later by [**FP**] for $d = 3$ and [**FFP**] for $d = 1$).

2. Complex hyperbolic space and isometries

We give a brief summary of key definitions and facts about complex hyperbolic space; see [**G**] and [**CG**] for more details (as well as [**KP**] for quaternionic hyperbolic space, which we will not discuss in detail here as the aspects that we consider are similar to the complex case). We will consider only the case of dimension $n = 2$ in this note, but the general setup is identical for higher dimensions so we state it for all $n \geqslant 1$. Consider $\mathbb{C}^{n,1}$, the vector space $\mathbb{C}^{n+1}$ endowed with a Hermitian form $\langle \cdot, \cdot \rangle$ of signature $(n, 1)$. Let $V^- = \{Z \in \mathbb{C}^{n,1} | \langle Z, Z \rangle < 0\}$. Let $\pi : \mathbb{C}^{n+1} - \{0\} \longrightarrow \mathbb{C}\mathrm{P}^n$ denote projectivization. Define $\mathrm{H}^n_{\mathbb{C}}$ to be $\pi(V^-) \subset \mathbb{C}\mathrm{P}^n$, endowed with the distance d (Bergman metric) given by:

$$\cosh^2 \frac{1}{2} d(\pi(X), \pi(Y)) = \frac{|\langle X, Y \rangle|^2}{\langle X, X \rangle \langle Y, Y \rangle} \tag{2.1}$$

From this formula it is clear that $\mathrm{PU}(n, 1)$ acts by isometries on $\mathrm{H}^n_{\mathbb{C}}$ (where $\mathrm{U}(n, 1)$ is the subgroup of $\mathrm{GL}(n+1, \mathbb{C})$ preserving $\langle \cdot, \cdot \rangle$, and $\mathrm{PU}(n, 1)$ is its image in $\mathrm{PGL}(n+1, \mathbb{C})$). The boundary at infinity $\partial_\infty \mathrm{H}^n_{\mathbb{C}}$ is naturally identified with $\pi(V^0) \subset \mathbb{C}\mathrm{P}^n$, where $V^0 = \{Z \in \mathbb{C}^{n,1} | \langle Z, Z \rangle = 0\}$.

FACT. $\mathrm{Isom}^0(\mathrm{H}^n_{\mathbb{C}}) = \mathrm{PU}(n, 1)$, and $\mathrm{Isom}(\mathrm{H}^n_{\mathbb{C}}) = \mathrm{PU}(n, 1) \ltimes \mathbb{Z}/2$ (complex conjugation).

CLASSIFICATION. $g \in \mathrm{PU}(2, 1) \setminus \{\mathrm{Id}\}$ is of one of the following types:

- *elliptic*: g has a fixed point in $\mathrm{H}^n_{\mathbb{C}}$
- *parabolic*: g has (no fixed point in $\mathrm{H}^n_{\mathbb{C}}$ and) exactly one fixed point in $\partial_\infty \mathrm{H}^n_{\mathbb{C}}$
- *loxodromic*: g has (no fixed point in $\mathrm{H}^n_{\mathbb{C}}$ and) exactly two fixed points in $\partial_\infty \mathrm{H}^n_{\mathbb{C}}$

DEFINITIONS. For any $1 \leqslant k \leqslant n$, a *complex k-plane* is a k-dimensional projective subspace of $\mathbb{C}P^n$ intersecting $\pi(V^-)$ non-trivially (so, it is an isometrically embedded copy of $\mathrm{H}^k_{\mathbb{C}} \subset \mathrm{H}^n_{\mathbb{C}}$). Complex 1-planes are usually called *complex lines*. A *complex reflection* is an elliptic isometry $g \in \mathrm{PU}(n, 1)$ whose fixed-point set is a complex $(n-1)$-plane. The eigenvalues of a matrix $A \in \mathrm{U}(n, 1)$ representing an elliptic isometry g have modulus one. Exactly one of these eigenvalues has eigenvectors in V^- (projecting to a fixed point of g in $\mathrm{H}^n_{\mathbb{C}}$), and such an eigenvalue will be called *of negative type*. An elliptic isometry $g \in \mathrm{PU}(n, 1)$ is a complex reflection if and only if the negative type eigenvalue of any of its matrix representatives has multiplicity n.

3. Picard modular groups and complex reflections

We use the *Siegel model* of hyperbolic space $\mathrm{H}^2_{\mathbb{C}}$, which is the projective model associated to the Hermitian form on $\mathbb{C}^3$ given by $\langle Z, W \rangle = W^* J Z$ with:

$$J = \begin{pmatrix} 0 & 0 & 1 \\ 0 & 1 & 0 \\ 1 & 0 & 0 \end{pmatrix}$$

The *Picard modular groups* are the arithmetic lattices $\Gamma_d = \mathrm{PU}(2, 1, \mathcal{O}_d)$ in $\mathrm{PU}(2, 1)$, where d is a squarefree positive integer and $\mathcal{O}_d$ is the ring of integers of

$\mathbb{Q}[i\sqrt{d}]$. The following elements belong to Γ_d for all d (taking $u = i$ when $d = 1$, $u = e^{i\pi/3}$ when $d = 3$, and $u = -1$ for all other values of d):

$$I_0 = \begin{bmatrix} 0 & 0 & 1 \\ 0 & -1 & 0 \\ 1 & 0 & 0 \end{bmatrix}, \qquad R = \begin{bmatrix} 1 & 0 & 0 \\ 0 & u & 0 \\ 0 & 0 & 1 \end{bmatrix}.$$

LEMMA 3.1. *I_0 and R are complex reflections.*

PROOF. R visibly has eigenvalues $\{1, 1, u\}$ and 1 is of negative type, hence R is a complex reflection. Likewise, I_0 has eigenvalues $\{-1, -1, 1\}$ with -1 of negative type (its eigenspace is the span of e_2 and $e_1 - e_3$), hence I_0 is a complex reflection. Note that I_0 has order 2, and R has order 2 except when $d = 3$ (when it has order 6) and $d = 1$ (when it has order 4). □

Given elements $\gamma_1, \ldots, \gamma_k$ of a group Γ, we denote $\langle\langle \gamma_1, \ldots, \gamma_k \rangle\rangle$ the normal closure of $\gamma_1, \ldots, \gamma_k$ in Γ, that is the smallest normal subgroup of Γ containing $\gamma_1, \ldots, \gamma_k$.

PROPOSITION 3.2. *Denote as above $\Gamma_d = \mathrm{PU}(2, 1, \mathcal{O}_d)$.*

- *When $d = 3$, $\langle\langle R \rangle\rangle = \Gamma_3$.*
- *When $d = 1$, $\langle\langle R, I_0 \rangle\rangle = \Gamma_1$, $\langle\langle I_0 \rangle\rangle$ has index 4 in Γ_1 and $\langle\langle R \rangle\rangle$ has index 96 in Γ_1.*
- *When $d = 7$, $\langle\langle I_0 \rangle\rangle = \Gamma_7$ and $\langle\langle R \rangle\rangle$ has index 168 in Γ_7.*
- *When $d = 2$, $\langle\langle R, I_0 \rangle\rangle = \langle\langle I_0 \rangle\rangle$ has index 4 in Γ_2.*
- *When $d = 11$, $\langle\langle R, I_0 \rangle\rangle$ has index 4 in Γ_{11} and $\langle\langle R \rangle\rangle$ has index 13,824 in Γ_{11}.*

PROOF. Let Γ be a group with (say, finite) presentation $\langle s_1, \ldots, s_n \,|\, r_1, \ldots, r_p \rangle$ and $w_1, \ldots, w_k$ elements of Γ given as words in the generators $s_1, \ldots, s_n$. Then, by construction of group presentations:

$$\hat{\Gamma} = \langle s_1, \ldots, s_n \,|\, r_1, \ldots, r_p, w_1, \ldots, w_k \rangle \simeq \Gamma / \langle\langle w_1, \ldots, w_k \rangle\rangle.$$

In particular, the order of $\hat{\Gamma}$ is equal to the index of $\langle\langle w_1, \ldots, w_k \rangle\rangle$ in Γ. The statements in the proposition give the results of this procedure applied to the presentations for the Γ_d obtained in [**MP**] and [**Po**], available as Magma files at [**MCode**] and [**PoCode**]. (Note that both R and I_0 appear conveniently as generators in these presentations.) More specifically, we add the relation R (resp. I_0, resp. R and I_0) to these presentations and use the Magma command `Order(G);` to compute the order of the quotient. Most of these computations can be done in a matter of seconds with the online Magma Calculator available at [**Mag**] (except for the higher orders 96, 168 and 13,824 which require an installed version of Magma).

To illustrate the procedure we apply it by hand to the remarkably simple presentation for Γ_3 obtained by Falbel–Parker (Theorem 5.9 of [**FP**]):

$$\Gamma_3 = \langle P, Q, R \,|\, R^2 = (QP^{-1})^6 = PQ^{-1}RQP^{-1}R = P^3Q^{-2} = (RP)^3 = 1 \rangle.$$

Note that we have an unfortunate conflict with the notation from [**FP**], as our "I_0" is their "R" and our "R" is their PQ^{-1}. Following here their notation, we see that:

$$\Gamma_3 / \langle\langle PQ^{-1} \rangle\rangle = \langle P, Q, R \,|\, P = Q, P^3 = Q^2, R^3 = R^2 = 1 \rangle$$

is the trivial group, whereas:

$$\Gamma_3/\langle\langle R\rangle\rangle = \langle P, Q \mid P^3 = Q^2 = (QP^{-1})^6 = 1\rangle$$

is a $(2,3,6)$ (Euclidean) triangle group, hence infinite. □

4. The Hurwitz modular group and quaternionic reflections

Quaternionic hyperbolic space admits a Siegel model analogous to the one discussed above for complex hyperbolic space, with the usual caveats of linear algebra over the quaternions (eg. matrices act on vectors by left multiplication and scalars act on vectors by right multiplication). See [**KP**] or [**MP**] for details. A *quaternionic reflection* is an elliptic isometry $g \in \mathrm{PSp}(n,1)$ whose fixed-point set is a quaternionic $(n-1)$-plane. We are interested in the lattice $\Gamma_{\mathcal{H}} = \mathrm{PSp}(2,1,\mathcal{H}) < \mathrm{PSp}(2,1)$, consisting of the (projectivized) matrices in $\mathrm{PSp}(2,1)$ whose entries lie in the Hurwitz ring $\mathcal{H} = \mathbb{Z}[i,j,k,\sigma] \subset \mathbb{H}$, denoting $\sigma = \frac{1-i+j+k}{2}$. The relevant elements of $\Gamma_{\mathcal{H}}$ for our purposes are:

$$I_0 = \begin{bmatrix} 0 & 0 & 1 \\ 0 & -1 & 0 \\ 1 & 0 & 0 \end{bmatrix}, \quad R_i = \begin{bmatrix} 1 & 0 & 0 \\ 0 & i & 0 \\ 0 & 0 & 1 \end{bmatrix}, \quad R_\sigma = \begin{bmatrix} 1 & 0 & 0 \\ 0 & \sigma & 0 \\ 0 & 0 & 1 \end{bmatrix}.$$

The same computation as the proof of Lemma 1 gives the following:

LEMMA 4.1. *I_0, R_i and R_σ are quaternionic reflections.*

PROPOSITION 4.2. *Let $\Gamma_{\mathcal{H}}$ and $I_0, R_i, R_\sigma \in \Gamma_{\mathcal{H}}$ as above. Then $\langle\langle R_\sigma\rangle\rangle = \Gamma_{\mathcal{H}}$, $\langle\langle I_0\rangle\rangle$ has index dividing 12 in $\Gamma_{\mathcal{H}}$ and $\langle\langle R_i\rangle\rangle$ has index dividing 648 in $\Gamma_{\mathcal{H}}$.*

PROOF. The procedure is the same as for the proof of Proposition 1, the difference being that we only use a partial presentation for $\Gamma_{\mathcal{H}}$, that is a presentation with all generators but only some of the relations. As observed in [**MP**], the presentation obtained there for $\Gamma_{\mathcal{H}}$ is too large for Magma to handle directly (it has 33 generators and 968,480 relations, and the text file is a bit over 200 MB). Rather, we use the partial presentation obtained by keeping only the 1000 first relations – a text file for this presentation is available as `QuaternionsTruncated1000.txt` at [**MCode**].

Denoting $\tilde{\Gamma}$ the abstract group with this partial presentation, we have a surjective homomorphism $\pi : \tilde{\Gamma} \longrightarrow \Gamma_{\mathcal{H}}$, obtained by adding the remaining relations for $\Gamma_{\mathcal{H}}$. Given a normal subgroup $H \trianglelefteq \tilde{\Gamma}$, π induces a surjective homomorphism $\tilde{\Gamma}/H \longrightarrow \Gamma_{\mathcal{H}}/\pi(H)$. Taking $H = \langle\langle R_\sigma\rangle\rangle$, $\langle\langle I_0\rangle\rangle$, $\langle\langle R_i\rangle\rangle$ successively, we compute the order of $\tilde{\Gamma}/H$ as above by adding the single relation R_σ, I_0, R_i to the presentation for $G = \tilde{\Gamma}$ using the Magma command `Order(G);` and the result follows. □

Acknowledgment

The authors would like to thank Matthew Stover for suggesting the direct computational method used in this note to determine indices of subgroups generated by reflections.

References

[Ag] Ian Agol, *Finiteness of arithmetic Kleinian reflection groups*, International Congress of Mathematicians. Vol. II, Eur. Math. Soc., Zürich, 2006, pp. 951–960. MR2275630

[Al1] Daniel Allcock, *New complex- and quaternion-hyperbolic reflection groups*, Duke Math. J. **103** (2000), no. 2, 303–333, DOI 10.1215/S0012-7094-00-10326-2. MR1760630

[Al2] Daniel Allcock, *The Leech lattice and complex hyperbolic reflections*, Invent. Math. **140** (2000), no. 2, 283–301, DOI 10.1007/s002220050363. MR1756997

[BFMS] Uri Bader, David Fisher, Nicholas Miller, and Matthew Stover, *Arithmeticity, superrigidity, and totally geodesic submanifolds*, Ann. of Math. (2) **193** (2021), no. 3, 837–861, DOI 10.4007/annals.2021.193.3.4. MR4250391

[Bi] Luigi Bianchi, *Sui gruppi di sostituzioni lineari con coefficienti appartenenti a corpi quadratici immaginarî* (Italian), Math. Ann. **40** (1892), no. 3, 332–412, DOI 10.1007/BF01443558. MR1510727

[BeMc] Mikhail Belolipetsky and John Mcleod, *Reflective and quasi-reflective Bianchi groups*, Transform. Groups **18** (2013), no. 4, 971–994, DOI 10.1007/s00031-013-9245-6. MR3127984

[CS] Donald I. Cartwright and Tim Steger, *Enumeration of the 50 fake projective planes* (English, with English and French summaries), C. R. Math. Acad. Sci. Paris **348** (2010), no. 1-2, 11–13, DOI 10.1016/j.crma.2009.11.016. MR2586735

[CG] S. S. Chen and L. Greenberg, *Hyperbolic spaces*, Contributions to analysis (a collection of papers dedicated to Lipman Bers), Academic Press, New York, 1974, pp. 49–87. MR0377765

[DM] P. Deligne and G. D. Mostow, *Monodromy of hypergeometric functions and nonlattice integral monodromy*, Inst. Hautes Études Sci. Publ. Math. **63** (1986), 5–89. MR849651

[D] Martin Deraux, *A new nonarithmetic lattice in* PU(3,1), Algebr. Geom. Topol. **20** (2020), no. 2, 925–963, DOI 10.2140/agt.2020.20.925. MR4092315

[DPP] Martin Deraux, John R. Parker, and Julien Paupert, *New nonarithmetic complex hyperbolic lattices II*, Michigan Math. J. **70** (2021), no. 1, 135–205, DOI 10.1307/mmj/1592532044. MR4255091

[FFP] Elisha Falbel, Gábor Francsics, and John R. Parker, *The geometry of the Gauss-Picard modular group*, Math. Ann. **349** (2011), no. 2, 459–508, DOI 10.1007/s00208-010-0515-5. MR2753829

[FP] Elisha Falbel and John R. Parker, *The geometry of the Eisenstein-Picard modular group*, Duke Math. J. **131** (2006), no. 2, 249–289, DOI 10.1215/S0012-7094-06-13123-X. MR2219242

[G] William M. Goldman, *Complex hyperbolic geometry*, Oxford Mathematical Monographs, The Clarendon Press, Oxford University Press, New York, 1999. MR1695450

[KP] Inkang Kim and John R. Parker, *Geometry of quaternionic hyperbolic manifolds*, Math. Proc. Cambridge Philos. Soc. **135** (2003), no. 2, 291–320, DOI 10.1017/S030500410300687X. MR2006066

[K] Bruno Klingler, *Sur la rigidité de certains groupes fondamentaux, l'arithméticité des réseaux hyperboliques complexes, et les "faux plans projectifs"* (French, with English summary), Invent. Math. **153** (2003), no. 1, 105–143, DOI 10.1007/s00222-002-0283-2. MR1990668

[Mag] Magma Computational Algebra System. Available at: `http://magma.maths.usyd.edu.au/magma/`

[MCode] A. Mark; companion files for [MP], available at `https://github.com/alice-mark/LatticePresentations`.

[MP] A. Mark and J. Paupert; *Presentations for cusped arithmetic hyperbolic lattices*, To appear in Algebr. Geom. Topol.

[Mos] G. D. Mostow, *Generalized Picard lattices arising from half-integral conditions*, Inst. Hautes Études Sci. Publ. Math. **63** (1986), 91–106. MR849652

[PWi] Julien Paupert and Pierre Will, *Real reflections, commutators, and cross-ratios in complex hyperbolic space*, Groups Geom. Dyn. **11** (2017), no. 1, 311–352, DOI 10.4171/GGD/398. MR3641843

[Po] David Polletta, *Presentations for the Euclidean Picard modular groups*, Geom. Dedicata **210** (2021), 1–26, DOI 10.1007/s10711-020-00531-9. MR4200907

[PoCode] D. Polletta; companion files for [**Po**], available at `https://github.com/DPolletta/Code-for-Euclidean-Picard-modular-group-derivations`.

[PY] Gopal Prasad and Sai-Kee Yeung, *Fake projective planes*, Invent. Math. **168** (2007), no. 2, 321–370, DOI 10.1007/s00222-007-0034-5. MR2289867

[Sh] M. K. Shaiheev, *Reflective subgroups in Bianchi groups*, Selecta Math. Soviet. **9** (1990), no. 4, 315–322. Selected translations. MR1078260

[St] Matthew Stover, *Arithmeticity of complex hyperbolic triangle groups*, Pacific J. Math. **257** (2012), no. 1, 243–256, DOI 10.2140/pjm.2012.257.243. MR2948468

Department of Mathematics, Vanderbilt University
Email address: `alice.h.mark@vanderbilt.edu`

School of Mathematical and Statistical Sciences, Arizona State University
Email address: `paupert@asu.edu`

School of Mathematical and Statistical Sciences, Arizona State University
Email address: `dpollett@asu.edu`

Contemporary Mathematics
Volume **783**, 2023
https://doi.org/10.1090/conm/783/15700

Verifying the Straight-and-spaced Condition

J. Maxwell Riestenberg

ABSTRACT. A classical local-to-global principle for undistorted actions on negatively curved spaces has been generalized to higher rank symmetric spaces by Kapovich-Leeb-Porti; this condition is equivalent to the Anosov property defined by Labourie and Guichard-Wienhard. In previous work the author found explicit criteria for this higher rank condition which, in principle, can be verified on a finite subset of the Cayley graph. In this note, we demonstrate that while the classical version can be checked, it is impractical to verify the higher-rank version with currently available techniques. Nonetheless, the numerical experiments here may give insight towards future work on this problem.

Introduction

This manuscript addresses one approach to the following problem:

PROBLEM 0.1. Given a representation of a hyperbolic group $\rho\colon \Gamma \to G$ into a semisimple Lie group, determine if ρ has finite kernel and discrete image.

Kapovich [**Kap16**] has shown that the discreteness problem is undecidable already for nonelementary subgroups of $\mathrm{PSL}(2,\mathbb{C})$. However, one can attempt to answer this question by proving a stronger condition, e.g. that the orbit map $\Gamma \to \mathbb{H}^n$ given by $\gamma \mapsto \rho(\gamma)p$ is a quasi-isometric embedding. It is possible to verify this condition with a finite check in the Cayley graph of Γ thanks to, for example, Theorem 1.2 below. Kapovich-Leeb-Porti proved that an analogue of Theorem 1.2 holds in higher rank symmetric spaces [**KLP14**], and the author found explicit criteria for their higher rank version [**Rie21**]. In this manuscript, we check the classical version in negative curvature and compare it to the higher rank version. A naive implementation involves enumerating words of length $3N$ and performing a check on certain combinations of words of length N. This approach fails the classical straight-and-spaced condition when $N = 3$. A slightly different approach involves enumerating words of length $2N$ and performing a check on certain pairs of words of length N. This version satisfies the classical straight-and-spaced condition when $N = 4$ but still badly fails the higher rank version. Nonetheless, we observe that the midpoint sequences in this second case satisfy a nice geometric condition, explained

2020 *Mathematics Subject Classification.* Primary 53C35; Secondary 20F65, 22E40.
Key words and phrases. Differential geometry, Geometric topology.
This work was partially funded by the Deutsche Forschungsgemeinschaft (DFG, German Research Foundation) under Germany's Excellence Strategy EXC 2181/1 - 390900948 (the Heidelberg STRUCTURES Excellence Cluster).

further in Section 3. This suggests that a practical version of the Kapovich-Leeb-Porti algorithm may still be feasible if certain estimates in [**Rie21**] can be improved.

1. Background

We let $\mathbb{H}^2$ denote the hyperbolic plane, normalized to have sectional curvature -1. In the code, we use the hyperboloid model of $\mathbb{H}^2$.

DEFINITION 1.1. Let (x_n) be a sequence of points in $\mathbb{H}^2$.

(1) The sequence is a (c_1, c_2)-*lower quasigeodesic* for $c_1 \geq 1$ and $c_2 \geq 0$ if for all m, n we have
$$\frac{1}{c_1}|m-n| - c_2 \leq d(x_n, x_m).$$
(2) The sequence is s-*spaced* for $s \geq 0$ if for all n we have
$$d(x_n, x_{n+1}) \geq s.$$
(3) The sequence is θ-*straight* for $0 \leq \theta < \pi$ if for all n,
$$\angle_{x_n}(x_{n-1}, x_{n+1}) \geq \pi - \theta.$$

Sufficiently straight-and-spaced sequences of points in a negatively curved space are known to be lower-quasigeodesics:

THEOREM 1.2 ([**KL19, Kap21**]). *Let (x_n) be an s-spaced and θ-straight sequence in $\mathbb{H}^d$. If there exists $\nu > 1$ such that*
$$\cosh\left(\frac{s}{2}\right)\cos\left(\frac{\theta}{2}\right) \geq \nu$$
then $n \mapsto x_n$ is a lower-quasigeodesic.

REMARK 1.3. Note that the definition of θ-straight here is slightly different than the definition given in [**Kap21**]. As a result, the statement here has $\cos\left(\frac{\theta}{2}\right)$ appear instead of $\sin\left(\frac{\theta}{2}\right)$. The convention here agrees with [**KLP14, Rie21**].

Kapovich-Leeb-Porti proved that an analogue of Theorem 1.2 holds in higher rank symmetric spaces [**KLP14**], and the author found explicit criteria for the higher rank version [**Rie21**, Theorem 5.1]. The subgroup $\mathrm{SO}(2,1) \subset \mathrm{SL}(3,\mathbb{R})$ acts on a totally geodesic copy of the hyperbolic plane in the symmetric space $\mathrm{SL}(3,\mathbb{R})/\mathrm{SO}(3)$. One criterion of [**Rie21**, Theorem 5.1] is a straightness condition on ζ-angles, which agree with Riemannian angles for points in this hyperbolic plane. The experiments here demonstrate that this example badly fails the higher rank straightness condition.

2. Numerical experiments

We choose an explicit surface subgroup Γ of $\mathrm{SO}(2,1)$ and attempt to verify the straight-and-spaced criteria. The group we consider is generated by
$$\left\{ \begin{bmatrix} \cos\theta & \sin\theta & 0 \\ -\sin\theta & \cos\theta & 0 \\ 0 & 0 & 1 \end{bmatrix} \begin{bmatrix} \cosh\lambda & 0 & \sinh\lambda \\ 0 & 1 & 0 \\ \sinh\lambda & 0 & \cosh\lambda \end{bmatrix} \begin{bmatrix} \cos\theta & -\sin\theta & 0 \\ \sin\theta & \cos\theta & 0 \\ 0 & 0 & 1 \end{bmatrix} \middle| \theta \in \left\{0, \frac{\pi}{8}, \frac{\pi}{4}, \frac{3\pi}{8}\right\} \right\}$$
for $\cosh(\frac{\lambda}{2}) = \cot(\frac{\pi}{8})$. This group is isomorphic to the fundamental group of a closed surface of genus 2. If we label the generators by a, b, c, d corresponding to $\theta = 0, \frac{\pi}{8}, \frac{\pi}{4}, \frac{3\pi}{8}$ respectively, then they satisfy the relation $adc^{-1}ba^{-1}d^{-1}cb^{-1} = 1$.

2.1. Consecutive triples of midpoints. In the first experiment, we use an enumeration of the geodesic words of length 9 in Γ.[1] We decompose each such word w into three words, each of length 3, $w = w_1 w_2 w_3$ and consider the midpoints

$$m_1 = \text{mid}(p, w_1 p), \quad m_2 = \text{mid}(w_1 p, w_1 w_2 p), \quad m_3 = \text{mid}(w_1 w_2 p, w_1 w_2 w_3 p).$$

The distance

$$s_w = \min(d(m_1, m_2), d(m_2, m_3))$$

gives a spacing parameter and the angle

$$\epsilon_w = \pi - \angle_{m_2}(m_1, m_2)$$

gives an angle parameter. Set

$$s_{min} = \min_{|w|=9} s_w$$

and

$$\epsilon_{max} = \max_{|w|=9} \epsilon_w.$$

Let (γ_n) be a geodesic sequence in Γ, i.e. a sequence of elements of Γ such that $d(\gamma_n, \gamma_m) = |n - m|$, where d is the word metric determined by the given generators. We consider the sequence of midpoints of $(\gamma_{3n} p)$ given by $m_k = \text{mid}(\gamma_{3k} p, \gamma_{3k+3} p)$. The sequence (m_n) is s_{min}-spaced and ϵ_{max}-straight. We check if our example passes the straight-and-spaced condition: Theorem 1.2 applies to the sequence of midpoints (m_n) if $\cos\left(\frac{\epsilon_{max}}{2}\right) > \operatorname{sech}\left(\frac{s_{min}}{2}\right)$ and this inequality holds if and only if $\cos(\angle_{m_2}(m_1, m_2)) < 4(1 + \cosh(s_{min}))^{-1} - 1 \approx -0.34$ for all words of length 9. Figure 1 demonstrates that our example fails this check.

2.2. Consecutive pairs of midpoints. The second experiment is similar to the first experiment, but we consider consecutive pairs of midpoints instead of consecutive triples of midpoints.

We use an enumeration of the geodesic words of length 8 in Γ. We decompose each word w into two words, each of length 4, $w = w_1 w_2$ and consider the midpoints

$$m_1 = \text{mid}(p, w_1^{-1} p), \quad m_2 = \text{mid}(p, w_2 p).$$

The distance

$$s_w = d(m_1, m_2)$$

gives a spacing parameter and the angle

$$\epsilon_w = \angle_{m_1}(p, m_2)$$

gives an angle parameter. Set

$$s_{min} = \min_{|w|=8} s_w$$

and

$$\epsilon_{max} = 2 \max_{|w|=8} \epsilon_w.$$

For a geodesic sequence (γ_n) in Γ, we consider the sequence of midpoints of $(\gamma_{4n} p)$ given by $m_k = \text{mid}(\gamma_{4k} p, \gamma_{4k+4} p)$. The sequence (m_n) is s_{min}-spaced and ϵ_{max}-straight. We check if our example passes the straight-and-spaced condition: we find that $s_{min} \approx 3.08$ so we need to check if $\cos\left(\frac{\epsilon_{max}}{2}\right) > \operatorname{sech}\left(\frac{s_{min}}{2}\right) \approx 0.41$. Figure 2 demonstrates that our example passes this check.

[1] The code is available at `https://github.com/MaxRiestenberg/KLP-algorithm`

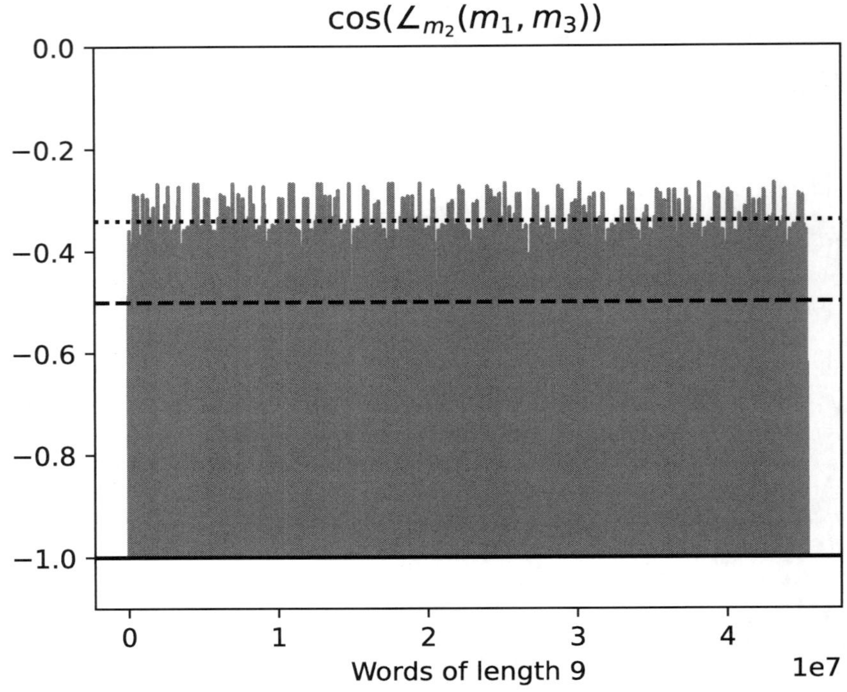

FIGURE 1. Since the plot does not stay below the dotted line $y = -0.34$, some consecutive triples of midpoints fail the straight-and-spaced check for hyperbolic space. This example also badly fails the higher rank check indicated by the bold line $y = -0.9997$. The significance of the dashed line $y = -0.5$ is explained Section 3.

3. The straight-and-spaced condition in higher rank

Theorem 1.2 has an analogue for sequences of points in higher rank symmetric spaces due to Kapovich-Leeb-Porti [**KLP14**]. The author found explicit criteria for their result in [**Rie21**]. The correct analogue for the straightness condition in higher rank is subtle and we do not review it here. Fortunately, it restricts to the usual Riemannian angle on totally geodesic copies of negatively curved spaces in higher rank symmetric spaces, so it is easy to compute in this example. Unfortunately, the straightness condition in higher rank is also much stricter, and the example we consider here badly fails the higher rank check which is indicated by the bold lines in Figures 1 and 2. This raises the question: can the estimates in [**Rie21**] be improved enough to obtain a practical algorithm?

An important technique in the higher rank local-to-global principle is to compare ζ-angles to distances to parallel sets [**KLP14**]; an explicit comparison is achieved in [**Rie21**, Lemma 4.16]. This comparison is only valid for ζ-angles very close to π and for this reason our example fails the higher rank check. The proof of [**Rie21**, Lemma 4.16] relies on a uniform estimate on the third derivative of

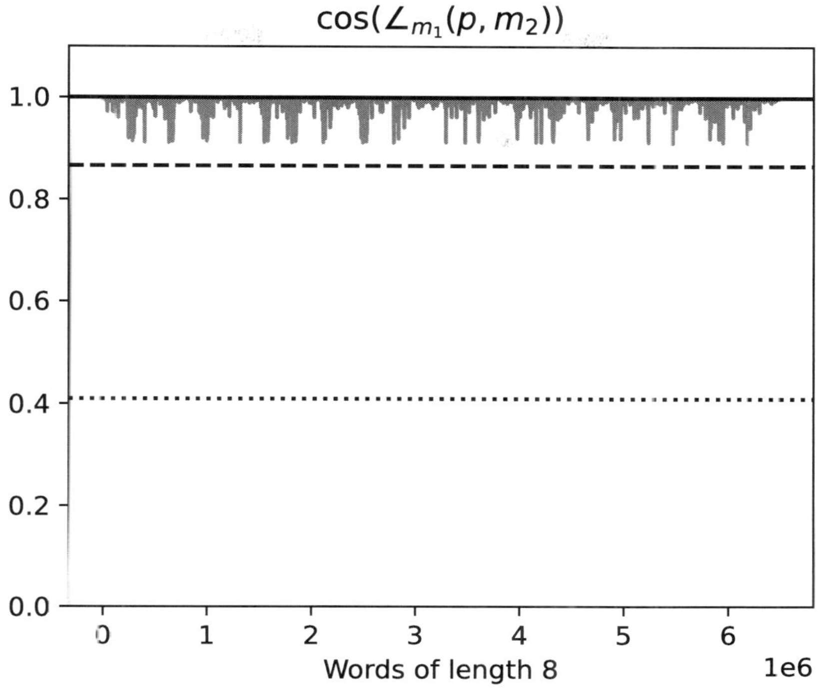

FIGURE 2. Since the plot stays above the dotted line $y = 0.41$, this example passes the check from hyperbolic space, but it badly fails the higher rank check indicated by the bold line $y = 0.99993$. The significance of the dashed line $y = 0.867$ is discussed below.

$f \circ \gamma \colon \mathbb{R} \to \mathbb{R}$ for any unit-speed geodesic γ in a fixed symmetric space and any Busemann function f. This estimate does not appear to be optimal for $\mathrm{SL}(3, \mathbb{R}) / \mathrm{SO}(3)$; it yields an upper bound of $2/3$ while numerical experiments suggest that the upper bound is close to 0.51. Unfortunately, even if we consider a hypothetical improved version of [**Rie21**, Lemma 4.16] using the upper bound of 0.51, the higher rank straightness condition is barely relaxed.

On the other hand, there is some reason to be optimistic that a practical implementation of the higher rank straight-and-spaced criteria is feasible. The proof that a straight and spaced sequence is a lower-quasigeodesic involves finding a pair of opposite Weyl cones V_+, V_- so that the sequence stays within a uniformly bounded distance of $V_- \cup V_+$. Candidates for these Weyl cones are provided by the ideal endpoints of the geodesic rays m_2m_1 and m_2m_3 for any consecutive triple m_1, m_2, m_3 in the straight and spaced sequence. However, to know that these candidate Weyl cones are actually opposite, we rely on the estimate $\angle^{\zeta}_{m_2}(m_1, m_3) > \pi - \varepsilon(\zeta)$. The analogous requirement in $\mathbb{H}^2$ is simply that the ideal endpoints are distinct. For $\mathrm{SL}(3, \mathbb{R}) / \mathrm{SO}(3)$, the check is $\angle^{\zeta}_{m_2}(m_1, m_3) > \pi - \varepsilon(\zeta) = \frac{2\pi}{3}$ for consecutive triples. This check is demonstrated in Figures 1 and 2 by the dashed line. Figure 1 shows that the angles between consecutive triples of midpoints of

$(\gamma_{3n}p)$ are not sufficiently straight to pass this check. However, Figure 2 shows that the corresponding condition on pairs of midpoints of $(\gamma_{4n}p)$, which is $\angle^{\zeta}_{m_1}(p, m_2) < \frac{1}{2}\varepsilon(\zeta) = \frac{\pi}{6}$, is satisfied. This suggests that the general framework of the Kapovich-Leeb-Porti algorithm may still lead to a practical implementation, and future efforts can reasonably focus on improving the estimates in [**Rie21**]; in particular in the control on distances to parallel sets in terms of ζ-angles.

Acknowledgments

The author is grateful to the organizers of the July 2021 ICERM virtual workshop *Computational Aspects of Discrete Subgroups of Lie Groups*: Alla Detinko, Alex Kontorovich, Peter Sarnak, Richard Schwartz and especially Michael Kapovich who proposed this work. The author also thanks Theodore Weisman who provided the list of geodesic words via the kbmag package in GAP [**GAP, kbmag**]. The numerical computations were performed in NumPy [**NumPy**].

References

[GAP] The GAP Group, *GAP – Groups, Algorithms, and Programming, Version 4.11.1*; 2021, `https://www.gap-system.org`

[GW12] Olivier Guichard and Anna Wienhard, *Anosov representations: domains of discontinuity and applications*, Invent. Math. **190** (2012), no. 2, 357–438, DOI 10.1007/s00222-012-0382-7. MR2981818

[Kap16] Michael Kapovich, *Discreteness is undecidable*, Internat. J. Algebra Comput. **26** (2016), no. 3, 467–472, DOI 10.1142/S0218196716500193. MR3506344

[Kap21] Michael Kapovich, *Geometric Algorithms for Discreteness and Faithfulness*, Preprint, 2021. `https://www.math.ucdavis.edu/~kapovich/EPR/hyperbolic-algorithm-2021-10-21.pdf`

[kbmag] Derek F. Holt, *kbmag - a GAP package, Version 1.5.9*; 2019, `https://gap-packages.github.io/kbmag/`

[KLP14] Michael Kapovich, Bernhard Leeb, and Joan Porti, *Morse actions of discrete groups on symmetric spaces*, Preprint, 2014. `arXiv:1403.7671`.

[KL19] Michael Kapovich and Beibei Liu, *Geometric finiteness in negatively pinched Hadamard manifolds*, Ann. Acad. Sci. Fenn. Math. **44** (2019), no. 2, 841–875, DOI 10.5186/aasfm.2019.4444. MR3973544

[Lab06] François Labourie, *Anosov flows, surface groups and curves in projective space*, Invent. Math. **165** (2006), no. 1, 51–114, DOI 10.1007/s00222-005-0487-3. MR2221137

[NumPy] Charles R. Harris, K. Jarrod Millman, Stéfan J. van der Walt, et al., *Array programming with NumPy*, Nature **585** (2020), pp. 357–362. DOI 10.1038/s41586-020-2649-2.

[Rie21] J. Maxwell Riestenberg, *A quantified local-to-global principle for Morse quasigeodesics*, Preprint, 2021. `arXiv:2101.07162`.

Cluster of Excellence STRUCTURES, Heidelberg University, 69120 Heidelberg, Germany

Email address: `mriestenberg@mathi.uni-heidelberg.de`

Contemporary Mathematics
Volume **783**, 2023
https://doi.org/10.1090/conm/783/15697

Unipotent Generators for Arithmetic Groups

T. N. Venkataramana

ABSTRACT. We sketch a simplification of proofs of old results on the arithmeticity of the group generated by opposing integral unipotent radicals contained in higher rank arithmetic groups.

1. Introduction

A well known theorem of Jacques Tits [**Tits1**] says that if $n \geq 3, k \geq 1$ are integers, then the group generated by upper and lower triangular unipotent matrices in the principal congruence subgroup $SL_n(\mathbb{Z}, k\mathbb{Z})$ of level k, has finite index in $SL_n(\mathbb{Z})$ (to save notation, we write $SL_n(k\mathbb{Z})$ for the group $SL_n(\mathbb{Z}, k\mathbb{Z})$ since this will be used for many kinds of principal congruence subgroups). This theorem admits a generalisation which will be described below (for definitions of the terms involved, see section 2).

Let $G \subset SL_n$ be a semi-simple $\mathbb{Q}$-simple algebraic group defined over $\mathbb{Q}$ and let $Q \subset G$ be a proper parabolic $\mathbb{Q}$-subgroup with unipotent radical U^+. Let U^- be the opposite unipotent radical and for an integer $k \geq 1$, denote by $E_Q(k)$ the subgroup generated by $U^+ \cap SL_n(k\mathbb{Z})$ and $U^- \cap SL_n(k\mathbb{Z})$. The aim of this note is to provide a proof (which is perhaps simpler and more uniform than the existing ones in the literature) of the following result.

THEOREM 1. *If $\mathbb{R}$-rank$(G) \geq 2$, then the group $E_Q(k)$ is an arithmetic subgroup of $G(\mathbb{Q})$, i.e. has finite index in $G(\mathbb{Z}) = G \cap SL_n(\mathbb{Z})$.*

REMARK. Every semi-simple $\mathbb{Q}$-simple algebraic group G is the group obtained by (Weil) restriction of scalars, of an absolutely simple algebraic group $\mathcal{G}$ defined over a number field K : $G = R_{K/\mathbb{Q}}(\mathcal{G})$. Theorem 1 is due to [**Tits1**] If $\mathcal{G}$ is a Chevalley group with K-*rank*$(\mathcal{G}) \geq 2$. For most of the classical groups, Theorem 1 was proved by Vaserstein [**Vaserstein1**], and [**Raghunathan1**] proved it for general groups of $\mathbb{Q}$-rank at least two. The remaining cases were proved in [**V1**].

These references prove Theorem 1 when Q is a *minimal* parabolic $\mathbb{Q}$- subgroup; however, as observed in [**V3**] and [**Oh2**], the general case follows easily from this case.

REMARK. A generalisation of Theorem 1 to the case when the arithmetic group $G(\mathbb{Z})$ is replaced by any Zariski dense discrete subgroup of $G(\mathbb{R})$ is proved in [**Oh1**],

2020 *Mathematics Subject Classification.* Primary 20H05; 14L15.

The support of JC Bose fellowship for the period 2021–2025 is gratefully acknowledged.

[**Benoist-Oh**], [**Benoist-Oh2**] and [**Benoist-Miquel**]; the proofs of these results are of a very different nature and we do not consider this situation. In fact, the proofs in these references *make use of* Theorem 1.

REMARK. The proof of Theorem 1 given here is uniform; however, this is based on Theorem 4 whose proof is not quite uniform but works especially well (see Section 3) when the group $G(\mathbb{R})$ is *not* a product of simple Lie groups of real rank one. In case $G(\mathbb{R})$ *is* a product of rank one groups, a more complicated argument is needed and we give the proof in Section 4.

We now describe another result from which Theorem 1 will be derived. Let G be a $\mathbb{Q}$-simple group with $\mathbb{Q}\text{-}rank(G) \geq 1$. Let $P \subset G$ be a proper *maximal* parabolic $\mathbb{Q}$ -subgroup. Denote by U^+ (resp. U^-) the unipotent radical (the "opposite unipotent radical") of P and let $P = LU^+$ be a Levi-decomposition of P; set $P^- = LU^-$, the parabolic subgroup of G "opposite" to P. Clearly L normalises $U^\pm$.

Denote by M the connected component of the Zariski closure of the group $L(\mathbb{Z})$ of integer points of L. The group M is non-trivial if and only if $\mathbb{R}\text{-}rank(G) \geq 2$ (Lemma 8). We denote by $V^\pm$ the commutator group $[M, U^\pm]$.

For each $k \geq 1$, denote by $F(k)$ the group generated by $V^\pm(k\mathbb{Z})$ and $M(k\mathbb{Z})$. Denote by $Cl(F(k))$ the closure of $F(k)$ in the group $G(\mathbb{A}_f)$ of finite adeles $\mathbb{A}_f$ over $\mathbb{Q}$. Let $\Gamma_k = G(\mathbb{Q}) \cap Cl(F(k))$. The group Γ_k has finite index in $G(\mathbb{Z})$ (Lemma 11); it is the smallest congruence subgroup of $G(\mathbb{Z})$ containing $F(k)$. We prove:

THEOREM 2. *If $\mathbb{R}\text{-}rank(G) \geq 2$, then $F(k)$ contains the commutator subgroup $[\Gamma_k, \Gamma_k]$:*

$$[\Gamma_k, \Gamma_k] \subset F(k).$$

We now show that Theorem 2 implies Theorem 1. By the Margulis normal subgroup theorem, the commutator $[\Gamma_k, \Gamma_k]$ has finite index in the higher rank arithmetic group Γ_k and is therefore an arithmetic group; therefore, by Theorem 2, the group

$$F(k) = < V^+(k), V^-(k), M(k) >$$

is an arithmetic group. Since M normalises V^+ and V^- it follows that $F(k)$ normalises the group $E'_P(k) = < V^+(k), V^-(k) >$ generated by $V^\pm(k)$. Therefore, again by the normal subgroup theorem, $E'_P(k)$ is an arithmetic group. Since $E'_P(k)$ is contained in the group $E_P(k) = < U^+(k), U^-(k) >$ it follows that the group $E_P(k)$ is an arithmetic group for every *maximal* parabolic $\mathbb{Q}$-subgroup P of G.

Let $Q \subset G$ be a parabolic $\mathbb{Q}$-subgroup as in Theorem 1. Fix a maximal parabolic $\mathbb{Q}$-subgroup P of G containing Q. Then $U_Q^\pm \supset U_P^\pm$. Hence $E_Q(k) = < U^+(k), U^-(k) >$ contains the group $E_P(k) = < U_P^+(k), U_P^-(k) >$. By the preceding paragraph, $E_P(k)$ is arithmetic and hence so is $E_Q(k)$; this proves Theorem 1.

Theorem 2 is deduced from a result on the centrality of the kernel for a map between two completions of the group $G(\mathbb{Q})$. This centrality is somewhat analogous to that of the centrality of the congruence subgroup kernel (in the case $\mathbb{R}\text{-}rank(G) \geq 2$), except that the congruence subgroup kernel is a compact (profinite) group. In our case, it is not clear, a priori, that C is even locally compact (it will follow after the fact that C is in fact finite). The details of the construction of the relevant completion will be given in Section 2. We briefly describe the construction here.

Equip the group $G(\mathbb{Q})$ with the topology $\mathcal{T}$ generated by the various cosets $\{gF(k) : g \in G(\mathbb{Q}), k \geq 1\}$ where $F(k)$ is as in Theorem 2. Then we prove in Section

2 the following proposition (note that even in the proposition the assumption of higher real rank is necessary).

PROPOSITION 3. *If* $\mathbb{R}$-*rank*$(G) \geq 2$, *then the group* G, *equipped with the topology* $\mathcal{T}$ *is a topological group.*

The topological group $(G, \mathcal{T})$ then admits a (two-sided) completion $\widehat{G}$ which can be shown to map onto $\overline{G} \subset G(\mathbb{A}_f)$ where $\overline{G}$ is the closure of $G(\mathbb{Q})$ in the finite adelic group $G(\mathbb{A}_f)$ (also referred to as the congruence completion of $G(\mathbb{Q})$); if G is simply connected, then by strong approximation, $\overline{G} = G(\mathbb{A}_f)$. Then we get an exact sequence

$$1 \to C \to \widehat{G} \to \overline{G} \to 1$$

of topological groups; the map $\widehat{G} \to \overline{G}$ can be shown to be an open map. The kernel C is closed in $\widehat{G}$; however, it is not clear a-priori that C is even compact). The main result of the paper is

THEOREM 4. *If* $\mathbb{R}$-*rank*$(G) \geq 2$, *then the kernel* C *is central in* $\widehat{G}$.

It can now be seen why this centrality implies Theorem 2. By the definition of the group Γ_k (as the smallest congruence subgroup containing $F(k)$), the groups $F(k)$ and Γ_k have the same closure in $G(\mathbb{A}_f)$. The openness of the map $\widehat{G} \to \overline{G}$ then implies that the closures $\widehat{\Gamma}_k, \widehat{F}(k)$ in $\widehat{G}$ of the groups $\Gamma_k, F(k)$ have the property that $\widehat{\Gamma}_k \subset C\widehat{F}(k)$. Therefore, by the centrality of C (Theorem 4) we see that

$$[\Gamma_k, \Gamma_k] \subset [\widehat{\Gamma}_k, \widehat{\Gamma}_k] \subset [\widehat{F}(k), \widehat{F}(k)] \subset \widehat{F}(k).$$

Therefore, we get

$$[\Gamma_k, \Gamma_k] \subset G(\mathbb{Q}) \cap \widehat{F}(k).$$

From Lemma 5 we have that $G(\mathbb{Q}) \cap \widehat{F}(k) = F(k)$; therefore we get $[\Gamma_k, \Gamma_k] \subset F(k)$, proving Theorem 2.

The centrality of C is deduced in Section 3 when the group M is not abelian; this is shown to be a simple consequence of strong approximation. When the group M is not abelian, the proof is more complicated and is is dealt with in Section 4; this involves the analogues of some results which are essentially proved (but stated only for the congruence subgroup kernel, instead of our group C) in [**V2**].

2. Preliminaries

2.1. Topological Groups. Let G be a group and $\mathcal{C} = \{W\}$ a (countable) collection of subgroups W with $\cap_{W \in \mathcal{C}} W = \{1\}$, and such that for any finite set $F \subset \mathcal{C}$ there exists a subgroup $V \in \mathcal{C}$ such that

$$V \subset \cap_{W \in F} W.$$

Let $\mathcal{T}$ be the topology on G generated by the cosets xW with $x \in G$ and $W \in \mathcal{C}$.

LEMMA 5. *The pair* $(G, \mathcal{T})$ *is a topological group if and only if for any* $x \in G$ *and* $W \in \mathcal{C}$ *there exists a subgroup* $V \in \mathcal{C}$ *such that* $xWx^{-1} \supset V$.

PROOF. Suppose for $x \in G$ and $W \in \mathcal{C}$, there exists $V \in \mathcal{C}$ such that $xWx^{-1} \supset V$. Let $x, y \in G$ and put $z = xy$; let $\mathcal{U}$ be a neighbourhood of z. There exists $W \in \mathcal{C}$ such that zW is a neighbourhood of z and $zW \subset \mathcal{U}$. By our assumptions on $\mathcal{C}$, there exists a $V \in \mathcal{C}$ such that $V \subset yWy^{-1} \cap W$. We then get

$$xVyV = xyy^{-1}VyV \subset xyWW = xyW = zW,$$

proving that the multiplication map $(x,y)\mapsto xy=z$ is continuous at (x,y). Moreover, $x^{-1}W\supset (Wx)^{-1}=(xx^{-1}Wx)^{-1}\supset (xV)^{-1}$ for some $V\in\mathcal{C}$, proving the continuity of $x\mapsto x^{-1}$. Therefore, $(G,\mathcal{T})$ is a topological group.

If the pair $(G,\mathcal{T})$ is a topological group, then the map $x\mapsto x^{-1}$ is continuous, and hence given $W\in\mathcal{C}$ and $x\in G$, the group xWx^{-1} is an open subgroup and hence contains some $V\in\mathcal{C}$. □

EXAMPLE. Take $G\subset SL_n(\mathbb{Q})$ to be a $\mathbb{Q}$-algebraic subgroup and $\mathcal{C}$ to be the collection, of principal congruence subgroups $G(k\mathbb{Z}):=G\cap SL_n(k\mathbb{Z})$ of integral matrices in $G\cap SL_n(\mathbb{Z})$ congruent to the identity matrix modulo k with $k\geq 1$. Then $(G,\mathcal{T})$ becomes a topological group and the topology on $G(\mathbb{Q})$ is the *congruence topology*.

REMARK. If the condition of Lemma 5 is satisfied, we say that a sequence $\{x_p\}$ of elements of G converges to an element y, if given a subgroup $W\in\mathcal{C}$, there exists an integer $p(W)$ such that for all $p\geq p(W)$, we have $x_p^{-1}y,\quad yx_p^{-1}\in W$.

2.2. Completions of Topological Groups. Let G be a topological group and $\mathcal{C}=\{W\}$ a collection of subgroups as in 2.1. We will say that a sequence $\{x_n\}_{n\geq 1}$ in G is a (two sided) Cauchy sequence if given a subgroup $W\in\mathcal{C}$, there exists an integer $n(W)$ such that

$$x_n^{-1}x_{n+m}\in W,\quad x_{n+m}x_n^{-1}\in W\quad (\forall\quad n\geq n(W),\quad \forall\quad m\geq 1).$$

We will say that two Cauchy sequences $\{x_n\},\{y_n\}$ are *equivalent* if given $W\in\mathcal{C}$, there exists an integer $n(W)$ such that

$$x_n^{-1}y_n\in W,\quad y_nx_n^{-1}\in W\quad\forall\quad n\geq n(W).$$

Denote by $\widehat{G}$ the set of equivalence classes of Cauchy sequences; elements of the original group G may be thought of as the set of constant Cauchy sequences. The resulting map $G\to\widehat{G}$ is an embedding (this follows from the assumption that the intersection $\cap W_{W\in\mathcal{C}}=\{1\}$). If $x=\{x_n\},y=\{y_n\}\in\widehat{G}$ denote by xy and x^{-1} respectively the sequences $\{x_ny_n\}$ and $\{x_n^{-1}\}$; it is routine to see that these sequences are Cauchy and we then get the structure of a group on $\widehat{G}$.

Given $W\in\mathcal{C}$, denote by $\widehat{W}$ the set of Cauchy sequences $x=(\{x_n\}$ such that for some integer $n(W)$ we have $x_n\in W\quad\forall\quad n\geq n(W)$. Then $\widehat{W}$ is a subgroup of $\widehat{G}$. Write $\widehat{\mathcal{C}}$ for the collection of sets $\{\widehat{W}:W\in\mathcal{C}\}$, and by $\widehat{\mathcal{T}}$ the topology on $\widehat{G}$ generated by the the collection of cosets $\{x\widehat{W}:x\in\widehat{G},\quad W\in\mathcal{C}\}$. Then $\widehat{G}$ and $\widehat{\mathcal{C}}$ satisfy the conditions of 2.1 and hence $(\widehat{G},\widehat{\mathcal{T}})$ is a topological group, referred to as the (two sided) *completion* of $(G,\mathcal{T})$ (see [**Bour**, Chapter III, Section 3, Exercise 6]). It follows from the definitions that $\widehat{G}$ is complete in the sense that Cauchy sequences in $(\widehat{G},\widehat{\mathcal{T}})$ converge to an element of $\widehat{G}$.

We first note an easy consequence of the definitions.

LEMMA 6. *If W,G is as before then the intersection $\widehat{W}\cap G=W$.*

PROOF. Let $g=(g,g,g,\cdots)\in G$ and suppose it lies in $\widehat{W}$. Therefore, there exists a Cauchy sequence $\{x_n\}_{n\geq 1}$ which is equivalent to the constant sequence g such that $x_n\in W$ for large enough n. Therefore, for m large enough, the elements $gx_m^{-1},x_m^{-1}g$ lie in W' for any given $W'\in\mathcal{C}$; in particular, if we take $W'=W$, we then see that for m large,

$$g = (gx_m^{-1})x_m \in W.W = W.$$

□

NOTATION. Given elements x, y of a group Γ, we write

$$^x(y) = xyx^{-1}, \quad [x, y] = xyx^{-1}y^{-1}.$$

If $\Delta \subset \Gamma$ is a subgroup, we write $^x(\Delta) = x\Delta x^{-1}$. If $A, B \subset \Gamma$ are subgroups, then $[A, B]$ denotes the subgroup generated by the commutators $[a, b]$ with $a \in A, b \in B$.

EXAMPLE. Let $G \subset SL_n$ be a linear algebraic group defined over $\mathbb{Q}$. Consider the collection $G(k\mathbb{Z}) = G \cap SL_n(k\mathbb{Z})$ of congruence subgroups in $G(\mathbb{Q})$. This collection satisfies the hypotheses of 2.1 and hence we get a completion denoted $\overline{G}$ of $G(\mathbb{Q})$, referred to as the *congruence completion* of $G(\mathbb{Q})$. This is also the closure of $G(\mathbb{Q})$ embedded as a subgroup of $G(\mathbb{A}_f)$ where $\mathbb{A}_f$ is the ring of finite adeles.

We may also consider the collection of subgroups of $G(\mathbb{Q})$ commensurable to $G(\mathbb{Z})$ (these are refereed to as *arithmetic subgroups* of $G(\mathbb{Q})$); the collection of arithmetic subgroups also satisfy the hypotheses of 2.1. We therefore get a completion $\widehat{G}_a$ of $G(\mathbb{Q})$, referred to as the *arithmetic completion* of $G(\mathbb{Q})$. We then have a surjective open map $\widehat{G}_a \to \overline{G}$ of topological groups split over $G(\mathbb{Q})$; the kernel C_G is seen to be a profinite (compact) group, called *the congruence subgroup kernel.*

The foregoing facts are well known ([**BMS**]).

2.3. Isotropic Algebraic Groups over $\mathbb{Q}$. In what follows, $G \subset SL_n$ is a $\mathbb{Q}$-simple linear algebraic group defined over $\mathbb{Q}$. It is said to be $\mathbb{Q}$-isotropic if there exists a torus $\mathbb{Q}$-isomorphic to the multiplicative group $\mathbb{G}_m$ embedded in G; let S in G be a maximal $\mathbb{Q}$-split torus and $\Phi = \Phi(\mathfrak{g}, S)$ be the roots (characters of S written additively) of S occurring in the Lie algebra $\mathfrak{g}$ of G under the adjoint action of S. Denote the *root space* of $\alpha \in \Phi$ by $\mathfrak{g}_\alpha$, the subspace of $\mathfrak{g}$ on which the torus S acts by the character $\alpha \in \Phi$. Fix a positive system of roots $\Phi^+ \subset \Phi$; then $\Phi = \Phi^+ \cup (-\Phi^+)$. We write $\alpha > 0$ if $\alpha \in \Phi^+$.

Denote by P_0 the connected subgroup of G whose Lie algebra is the direct sum $\mathfrak{g}^S \oplus_{\alpha \in \Phi^+} \mathfrak{g}_\alpha$ where $\mathfrak{g}^S$ denotes the subspace of vectors in $\mathfrak{g}$ fixed by the split torus S (the Lie algebra of the centraliser of S in G). Then P_0 is a minimal parabolic $\mathbb{Q}$-subgroup of G. Let $P \subset P_0$ be a *maximal* parabolic $\mathbb{Q}$-subgroup of G and U^+ its unipotent radical with Lie algebra $\mathfrak{u}^+$. Then $\mathfrak{u}^+ \subset \oplus_{\alpha>0}\mathfrak{g}_\alpha$ and is a sum $\mathfrak{u}^+ = \oplus_{\alpha \in X}\mathfrak{g}_\alpha$ of root spaces for some subset $X \subset \Phi^+$ of positive roots. There is a decomposition (the Levi decomposition) of P as a product $P = LU^+$, where L is a connected subgroup of G containing S, whose Lie algebra is the direct sum of the root spaces $\mathfrak{g}_{\pm\alpha}$ with $\alpha \notin X$ and $\mathfrak{g}^S$.

Let $\mathfrak{u}^- = \oplus_{\alpha \in X}\mathfrak{g}_{-\alpha}$ and U^- the connected (in fact unipotent) subgroup of G with Lie algebra U^-. This is called the *opposite* of U; the group $P^- = U^-L$ is a maximal parabolic $\mathbb{Q}$-subgroup called the *opposite of* P. The multiplication map $U^- \times P \to G$ given by $(v, p) \mapsto vp$ identifies the product space $U^- \times P$ as a Zariski dense open set $\mathcal{U} = U^-P$ in G defined over $\mathbb{Q}$. If $H \subset SL_n$ is a $\mathbb{Q}$-subgroup, and $\mathbb{Z}_p$ is the ring of p-adic integers, we write $H(k\mathbb{Z}_p)$ for the subgroup of elements of $H \cap SL_n(\mathbb{Z}_p)$ viewed as $n \times n$-matrices which are congruent to the identity matrix modulo k.

LEMMA 7. *Let $k \geq 1$ be an integer and for a prime p, consider the set $\mathcal{U}(k\mathbb{Z}_p) = U^-(k\mathbb{Z}_p)P(k\mathbb{Z}_p)$ where $\mathbb{Z}_p$ is the ring of p-adic integers. There exists a compact open subgroup $K_p(k)$ of $G(\mathbb{Z}_p)$ contained in $\mathcal{U}(k\mathbb{Z}_p)$.*

PROOF. The set $\mathcal{U}(k\mathbb{Z}_p)$ is an open subset of $G(\mathbb{Z}_p)$ containing 1. A fundamental system of neighbourhoods of identity in $G(\mathbb{Z}_p)$ is given by open subgroups and hence the lemma follows. □

2.4. The Groups M, V^+ and V^-. The group $L(\mathbb{Z})$ of integer points of the Levi subgroup L of subsection 2.3 is not Zariski dense in L (since the $\mathbb{Q}$-split central torus of L has only a finite number of integer points). Denote by M the connected component of the Zariski closure of $L(\mathbb{Z})$. The group M does not change if we replace $L(\mathbb{Z})$ by a finite index subgroup. Since $L(\mathbb{Q})$ commensurates $M(\mathbb{Z})$, it follows that its Zariski closure L normalises M.

LEMMA 8. *The dimension of M is positive if and only if $\mathbb{R}$-rank$(G) \geq 2$.*

PROOF. By definition, M is the connected component of identity of the Zariski closure of $L(\mathbb{Z})$; hence its dimension is zero if and only if $L(\mathbb{Z})$ is finite. Since L is the Levi subgroup of a maximal parabolic $\mathbb{Q}$- subgroup P of G, we may write $L = S_1L_1L_2$ as a product where S_1 is a one dimensional $\mathbb{Q}$-split torus, L_1 is the product of $\mathbb{Q}$-isotropic simple factors of L, and L_2 is a $\mathbb{Q}$-anisotropic group. (To see this, we write the semisimple part L^{ss} of L as a product L_1L_3 where L_1 is a product of $\mathbb{Q}$-isotropic simple groups L_1 and L_3 is a product of $\mathbb{Q}$-anisotropic simple groups; then $L = S_1T_1L^{ss}$ where S_1 is $\mathbb{Q}$ split torus in the centre of L, and T_1 is a $\mathbb{Q}$-anisotropic part of the centre of L. We may take $L_2 = L_3T_1$).

If $L(\mathbb{Z})$ is finite, then $L_2(\mathbb{Z})$ is finite and since L_2 is $\mathbb{Q}$-anisotropic, by the Godement criterion, $L_2(\mathbb{R})/L_2(\mathbb{Z})$ is compact and hence $L_2(\mathbb{R})$ is compact and has real rank zero. Since $L_1(\mathbb{Z})$ is also finite but L_1 is a product of $\mathbb{Q}$-isotropic $\mathbb{Q}$-simple groups, it follows that L_1 is trivial and hence $L = S_1L_2$ where L_2 has real rank 0. Consequently the real rank of L is the dimension of S_1 which is one and hence $\mathbb{R}$-$rank(G) = \mathbb{R}$-$rank(L) = 1$.

Conversely, if the real rank of G is one, then the group $L(\mathbb{R}) = S_1(\mathbb{R})L_1(\mathbb{R})L_2(\mathbb{R})$ has real rank one and hence L_1 is trivial and L_2 is anisotropic over $\mathbb{R}$; hence $L_2(\mathbb{R})$ is compact. Therefore, $L(\mathbb{Z}) \simeq S_1(\mathbb{Z})L_1(\mathbb{Z}) = \{\pm 1\}L_2(\mathbb{Z})$ is finite and hence M has dimension zero. □

Denote by $V^{\pm}$ the group $[M, U^{\pm}]$ generated by the commutators $mum^{-1}u^{-1}$ with $m \in M$ and $u \in U^{\pm}$. Since M is normal in L (and $U^{\pm}$ are normalised by L, it follows that $V^{\pm}$ is normalised by L. It is clear that $V^{\pm}$ are unipotent subgroups of $U^{\pm}$.

LEMMA 9. *Suppose G is $\mathbb{Q}$-simple and $\mathbb{Q}$-isotropic.*

[1] *The group $U^{\pm}$ normalises $V^{\pm}$.*

[2] *If $\mathbb{R}$-$rank(G) \geq 2$ then G is generated by the groups V^+, V^- and M.*

PROOF. The action of the reductive group M on the Lie algebra $\mathfrak{u}^{\pm}$ is completely reducible. Consequently, the lie algebra $\mathfrak{u}$ splits into the space $(\mathfrak{u}^{\pm})^M$ of M invariants and the space of *non-invariants* i.e. the span of $mXm^{-1} - X$ with $X \in \mathfrak{u}^{\pm}$ and $m \in M$. Since the non-invariants all lie in the Lie algebra $Lie(V^{\pm})$, it follows that $\mathfrak{u}^{\pm} = (\mathfrak{u}^{\pm})^M + (LieV)^{\pm}$ (it is possible that the Lie algebra *generated* by the non-invariants picks up invariant vectors; hence the sum may not be direct). If $X \in (\mathfrak{u}^{\pm})^M$, $m \in M$ and $Y \in \mathfrak{u}^{\pm}$, then we have

$$[X, m(Y) - Y] = [m(X), m(Y)] - [X, Y] = m([X, Y]) - [X, Y] \in (LieV)^{\pm}.$$

Therefore, $U^{\pm}$ normalises $(LieV)^{\pm}$ proving the first part.

Let G' be the group generated by $V^{\pm}$ and M; since all these groups are connected, so is G'; let $\mathfrak{g}'$ be its Lie algebra. We will show that $\mathfrak{g}$ normalises $\mathfrak{g}'$; the $\mathbb{Q}$-simplicity of G then implies that $\mathfrak{g}' = \mathfrak{g}$ and hence that $G' = G$. Since the group L normalises $U^{\pm}$ and also M, clearly L normalises $\mathfrak{g}'$ and hence its Lie algebra $\mathfrak{l}$ normalises $\mathfrak{g}'$. We therefore need to check that $\mathfrak{u}^{\pm}$ normalises $\mathfrak{g}'$. We first note that since M is (connected and) normal in M, we have $m(Z) - Z \in Lie(M) \subset \mathfrak{g}'$ if $Z \in \mathfrak{l}$. Since $[M, U^{\pm}] = V^{\pm}$, it follows that if $Z \in \mathfrak{u}^{\pm}$ then $m(Z) - Z \in LieV^{\pm} \subset \mathfrak{g}'$. Since the whole Lie algebra $\mathfrak{g}$ is spanned by $\mathfrak{u}^{\pm}$ and $\mathfrak{l}$, we have

$$m(Z) - Z \in \mathfrak{g}' \quad \forall \quad Z \in \mathfrak{g}. \tag{1}$$

We have proved in the proof of the first part of the lemma, that $\mathfrak{u}^{\pm} = (\mathfrak{u}^{\pm})^M + Lie(V^{\pm})$. The latter spaces $Lie(V^{\pm})$ are already contained in $\mathfrak{g}'$. Therefore, in order to verify that the sub-algebras $\mathfrak{u}^{\pm}$ normalise $\mathfrak{g}'$, it is enough to check that the M-invariants in $\mathfrak{u}^{\pm}$ normalise $\mathfrak{g}'$.

Suppose $X \in (\mathfrak{u}^{+})^M$ and $Y \in \mathfrak{u}^{\pm}$. Fix $m \in M$. We compute the bracket

$$[X, m(Y) - Y] = [X, m(Y)] - [X, Y] = [m(X), m(Y)] - [X, Y],$$

where the last equality follows because X is invariant under $m \in M$. Hence $[X, m(Y) - Y]$ is $m(Z) - Z$ with $Z = [X, Y]$. By equation (1), the bracket $[X, m(Y) - Y] = m(Z) - Z \in \mathfrak{g}'$. We have thus proved that $[(\mathfrak{u}^{+})^M, Lie(V^{\pm})] \subset \mathfrak{g}'$. If $Z \in Lie(M)$, then $[(\mathfrak{u}^{+})^M, Z] = 0 \subset \mathfrak{g}'$. Since $\mathfrak{g}'$ is generated by $Lie(V^{\pm})$ and $Lie(M)$ and each of these spaces, upon taking brackets with elements of $(\mathfrak{u}^{+})^M$ lie in $\mathfrak{g}'$, it follows that $[(\mathfrak{u}^{+})^M, \mathfrak{g}'] \subset \mathfrak{g}'$: the M-invariants in $\mathfrak{u}^{+}$ normalise $\mathfrak{g}'$. Similarly the M-invariants in $\mathfrak{u}^{-}$ also normalise $\mathfrak{g}'$. By the last remark of the preceding paragraph, $\mathfrak{u}^{\pm}$ normalises $\mathfrak{g}'$.

On the other hand $\mathfrak{l}$ is contained in the normaliser of $\mathfrak{g}'$. Therefore all of $\mathfrak{g}$ normalises $\mathfrak{g}'$ and the lemma follows. □

If $H \subset SL_n$ is an algebraic $\mathbb{Q}$-subgroup, we write $H(k\mathbb{Z})$ for the intersection $H \cap SL_n(k\mathbb{Z})$, where $SL_n(k\mathbb{Z})$ is the group of $n \times n$ matrices in $SL_n(\mathbb{Z})$ congruent to the identity matrix modulo k. Denote by $F(k)$ the group generated by $V^{\pm}(k\mathbb{Z})$ and $M(k\mathbb{Z})$. We note a corollary of lemma 9.

COROLLARY 1. *If $\mathbb{R}$-rank$(G) \geq 2$, then $F(k)$ is Zariski dense in G.*

PROOF. Since $V^{\pm}$ are unipotent $\mathbb{Q}$-groups, it is clear that $V^{\pm}(k\mathbb{Z})$ are Zariski dense in $V^{\pm}$. Moreover, $M(k\mathbb{Z}) \simeq L(k\mathbb{Z})$ is Zariski dense in M. Therefore, the Zariski closure of $F(k)$ contains $V^{\pm}$ and M. By Lemma 9, The Zariski closure of $F(k)$ is equal to G. □

2.5. Strong Approximation. We recall some well known results on strong approximation. Suppose $H \subset SL_n$ be a simply connected semi-simple $\mathbb{Q}$ -simple algebraic group with $\mathbb{R} - rank(G) \geq 1$. We work with the fixed embedding $H \subset SL_n$. Set $H(\mathbb{Z}) = H \cap SL_n(\mathbb{Z})$. Then strong approximation says that $H(\mathbb{Z})$ is dense in the group $H(\widehat{Z})$ where $\widehat{\mathbb{Z}} = \prod \mathbb{Z}_p$ where p runs through all primes. Let a, b be coprime integers and $H(a\mathbb{Z}) = H(\mathbb{Z}) \cap SL_n(a\mathbb{Z})$ where $SL_n(a\mathbb{Z})$ are integral matrices in $SL_n(\mathbb{Z})$ congruent to the identity matrix modulo the integer a. Then, as before, $H(a\mathbb{Z})$ is called the principal congruence subgroup of level a.

LEMMA 10. *If a, b are co-prime integers, and $H \subset SL_n$ is a simply connected $\mathbb{Q}$-simple with $\mathbb{R}$-rank$(H) \geq 1$, then $H(a\mathbb{Z})$ and $H(b\mathbb{Z})$ generate $H(\mathbb{Z})$.*

The same conclusion holds if H is semisimple, simply connected $\mathbb{Q}$ group which is product of $\mathbb{Q}$-simple (simply connected) groups H_i with $\mathbb{R}$-rank$(H_i) \geq 1$ for each i.

PROOF. This lemma and the proof are well known consequences of strong approximation. For the sake of completeness of the exposition, we recall the proof.

The definitions imply that $H(a\mathbb{Z}) = H(\mathbb{Z}) \cap H(a\widehat{\mathbb{Z}})$ is dense in $H(a\widehat{\mathbb{Z}})$. Thus the group Γ^* generated by the two principal congruence groups $H(a\mathbb{Z}), H(b\mathbb{Z})$ is also a congruence group, dense in the group H^* generated by $H(a\widehat{\mathbb{Z}}) = \prod_p H(a\mathbb{Z}_p)$ and $H(b\widehat{\mathbb{Z}}) = \prod_p H(b\mathbb{Z}_p)$; since a, b are coprime, *at each prime* p, the group generated by $H(a\mathbb{Z}_p)$ and $H(b\mathbb{Z}_p)$ is $H(\mathbb{Z}_p)$ and hence H^* is all of $H(\widehat{\mathbb{Z}})$. If two congruence subgroups of $H(\mathbb{Z})$ have the same closure in the congruence completion $H(\widehat{\mathbb{Z}})$, then they are the same; hence $\Gamma^* = H(\mathbb{Z})$ and first part of the lemma follows.

The second part readily follows from the first part applied to each H_i. □

LEMMA 11. *The intersection $\Gamma_k = G(\mathbb{Q}) \cap \overline{F}(k)$ where $\overline{F}(k)$ is the closure of the group $F(k)$ in the congruence completion $\overline{G}$ of $G(\mathbb{Q})$ is an arithmetic group (called the congruence closure of $F(k)$).*

PROOF. A theorem of Nori and Weisfeiler ([**Nori**], [**W**]) says, in particular, that if $\Gamma \subset G(\mathbb{Z})$ is a Zariski dense subgroup (and G is $\mathbb{Q}$-simple, $\mathbb{Q}$-isotropic), then the closure of Γ in the congruence completion (in this case $G(\mathbb{A}_f)$) is open). It is not difficult to extend this to the case when G is not necessarily simply connected (but the congruence completion $\overline{G}$ of G may not be all of $G(\mathbb{A}_f)$). Thus the intersection of $G(\mathbb{Q})$ with the closure of Γ is a congruence (arithmetic) subgroup of $G(\mathbb{Q})$; it is the smallest congruence subgroup of $G(\mathbb{Z})$ containing Γ and is called the congruence closure of Γ.

Applying this to the Zariski dense subgroup $F(k)$ (Corollary 1), we see that the congruence closure Γ_k of $F(k)$ has finite index in $G(\mathbb{Z})$ (in fact, in this case, one can prove directly by a somewhat lengthy argument that the closure of $F(k)$ in $\overline{G}$ is open, without using Nori-Weisfeiler). □

REMARK. Consider the group $P_F^{\pm} = MV^{\pm}$ where $V^{\pm} = [M, U^{\pm}]$. We have seen (Lemma 9) that $U^{\pm}$ normalises $V^{\pm}$; it then follows that $U^{\pm}$ normalises $MV^{\pm}$ as well, since

$$^{u}(mv) = [u, m]m^{u}(v) = m[^{m^{-1}}(u), m]^{u}(v) \in MV.$$

The groups M and $V^{\pm}$ are normalised also by L; hence $P^{\pm} = LU^{\pm}$ normalises $P_F^{\pm}$.. Since $P_F^{\pm}$ is the semi-direct product of the $\mathbb{Q}$ groups M and $V^{\pm}$, it follows from definitions that for varying integers k, $M(k)V^{\pm}(k)$ is a fundamental system of congruence subgroups of $P_F(\mathbb{Q})$; in particular given $p \in P(\mathbb{Q})$ and an integer $k \geq 1$, there exists an integer l such that $p(M(k)V^{\pm}(k))p^{-1} \supset M(l)V^{\pm}(l)$.

LEMMA 12. *Given a Zariski dense subset $D \subset U^{-}(\mathbb{Q})$ and an integer $k \geq 1$, there exists a finite set F in D and an integer $l \geq 1$ such that the group $M(l)V^{-}(l)$*

is contained in B where B is the group generated by the conjugates ${}^{v}(M(k)) := v(M(k))v^{-1}$ as v runs through elements of the finite set F.

PROOF. Fix an element $v^* \in D$. Consider the algebraic group V' generated by the elements of the form

$$\phi(u) ={}^{v^*} ([m^{-1}, u]) = [v^* m^{-1}(v^*)^{-1}]v^* umu^{-1}(v^*)^{-1} ={}^{v^*} ((m^{-1})^{u}(m))$$

with $u = (v^*)^{-1}v$ varying through the dense set $D' = (v^*)^{-1}D$ as v varies in D and m varies in M. Since D' is Zariski dense in U^-, it follows that this group V' is the group ${}^{v^*}([M, U^-] ={}^{v^*} (V^-) = V^-$ since U^- normalises V^-.

For reasons of dimension, there exists a finite set of these elements v such that the elements $\phi(u)$ generate a Zariski dense subgroup of the unipotent group U^- as m varies in $M(l)$ and v varies in F. Since these finite set of elements u are all rational, by choosing the congruence level l' suitably, we may assume that $\phi(u)$ are all elements in $U^-(k)$ for all $m \in M(l')$ and all $v \in F$. But a Zariski dense subgroup of integral elements in a unipotent group (namely V^-) contains $V^-(l'')$ for some congruence level l''. Moreover, since $\phi(u) = v^*(m)^v(m^{-1})$, the group ${}^{v^*}(M(k))$ together with $V^-(l'')$ generates a congruence subgroup containing $M(l)V^-(l)$ for some l. □

PROPOSITION 13. *Assume G is a $\mathbb{Q}$-simple $\mathbb{Q}$ isotropic algebraic group with $\mathbb{R} - rank(G) \geq 2$. Given $x \in G(\mathbb{Q})$ and $k \geq 1$, there exists an integer $l = l(k, x)$ such that*

$${}^{x}(F(k)) \supset F(l).$$

PROOF. For every $\theta \in F(k)$ we have ${}^{x}(F(k)) ={}^{x\theta} (F(k))$. Since $F(k)$ is Zariski dense in G, we may assume, by replacing x by $x\theta$ if necessary, that $x \in U^-P = \mathcal{U}$. Write $x = vp$ accordingly with $v \in U^-(\mathbb{Q})$ and $p \in P(\mathbb{Q})$ with $v = v(x)$ depending algebraically on $x \in \mathcal{U}$. Then,

$${}^{x}(F(k)) \supset^{x} (M(k)V^+(k)) \supset^{x} (M(k)V^+(k)) \cap M(k)V^-(k) =$$

$$\supset^{v} (M(l_1)V^+(l_1)) \cap M(k)V^-(k) \supset^{v} (M(l_1)V^+(l_1) \cap M(l_1)V^-(l_1)),$$

for some integer l_1 (since $v \in P^-(\mathbb{Q})$ normalises MV^-). Since $MV \cap MV^- = M$ we get:

$${}^{x}(F(k)) \supset^{v} (M(l_2))$$

for some integer l_2 with $v = v(x)$. Replacing x by any $x\gamma$ with $\gamma \in F(k)$, we see that ${}^{x}(F(k)) \supset^{v(x\gamma)} (M(l_\gamma))$ for some integer l_γ. By Lemma 12, for some finite set F of these γ's, the group generated by the conjugates

$${}^{v}(M(l_2)),{}^{v(x\gamma)} (M(l_\gamma)) \quad (\gamma \in F),$$

contains a congruence subgroup of the form $M(l\mathbb{Z})V^-(l)$ for some integer l and therefore, ${}^{x}(F(k)) \supset M(l)V^-(l)$ for some l; similarly, ${}^{x}(F(k)) \supset M(l)V^+(l)$ for some l and hence ${}^{x}(F(k))$ contains the group $F(l)$ generated by $V^{\pm}(l)$ and $M(l)$. □

2.6. The Group C. By (2.1) and by Proposition 13 we get the following. If G is a $\mathbb{Q}$-isotropic $\mathbb{Q}$-simple algebraic group with $\mathbb{R}$-*rank*$(G) \geq 2$, denote by $\mathcal{T}$ the topology on $G(\mathbb{Q})$ generated by the cosets $xF(k) : x \in G(\mathbb{Q}), k \geq 1$. Then $(G(\mathbb{Q}), \mathcal{T})$ gets the structure of a topological group. By (2.2) The topological group $(G, \mathcal{T})$ admits a two sided completion $(\widehat{G}, \widehat{\mathcal{T}})$. If, as before, $\overline{G} \subset G(\mathbb{A}_f)$ denotes the *congruence completion* of $G(\mathbb{Q})$, we get a surjective homomorphism $\widehat{G} \to \overline{G}$. This proves Proposition 3.

Since the group $F(k)$ lies in $G(k\mathbb{Z})$ the principal congruence subgroup of $G(\mathbb{Z}) = G \cap SL_n(\mathbb{Z})$ of level k, and since the $G(k\mathbb{Z}) : k \geq 1$ form a fundamental system of neighbourhoods of identity, it follows that any congruence subgroup of $G(\mathbb{Q})$ contains $G(k\mathbb{Z})$ for some k and hence contains $F(k)$. Since the group Γ_k is the smallest congruence subgroup of $G(\mathbb{Q})$ containing $F(k)$, it follows that the Γ_k form a fundamental system of neighbourhoods of identity in $G(\mathbb{Q})$ for the congruence topology. Since $F(k)$ is dense in n Γ_k, it follows that if l is a multiple of k, then the quotient set $F(k)/F(l)$ maps onto the *finite* congruence quotient set Γ_k/Γ_l. Taking inverse limits, it follows that $\widehat{F}(k)$ maps onto $Cl(\Gamma_k) = Cl(F(k))$ and the latter is an open subgroup of $\overline{G}$. Thus the map $\widehat{G} \to \overline{G}$ is an *open map*, with kernel C, say.

The kernel C is the inverse image of the completions $\widehat{\Gamma}_k$ as k varies. Moreover, the inverse limit of $\widehat{F}(k)$ is trivial. Hence we get

$$C = \varprojlim \widehat{F}(k)\backslash\widehat{\Gamma}_k/\widehat{F}(k) = \varprojlim F(k)\backslash\Gamma_k/F(k). \tag{2}$$

Note that the group $M(\mathbb{Z})$ normalises $V^{\pm}(k\mathbb{Z})$ and $M(k\mathbb{Z})$ and hence normalises $F(k), \Gamma_k$. Since C is normal in $\widehat{G}$, it is normalised by $G(\mathbb{Q}) \supset M(\mathbb{Z})$; the above expression (2) of C as the inverse limit of the double cosets $F(k)\backslash\Gamma_k/F(k)$ respects this $M(\mathbb{Z})$ action.

3. The Case Where M is not abelian

We now prove the centrality of the kernel C (Theorem 4) in the case when M is not abelian. Since M is connected reductive and is (the connected component of identity of) the Zariski closure of $L(\mathbb{Z})$, it follows that $M(\mathbb{Z})$ is Zariski dense in M; hence the commutator subgroup $S = [M, M]$ is a (non-trivial) semi-simple $\mathbb{Q}$-group with $S(\mathbb{Z})$ being Zariski dense in S. Let S^* denote the simply connected cover of S. Let $S^* = \prod S_i$ be a product of $\mathbb{Q}$-simple groups S_i. Since $S^*(\mathbb{Z})$ is Zariski dense in S^*, we have that $S_i(\mathbb{Z})$ is Zariski dense in each S_i, and hence each $S_i(\mathbb{R})$ is non-compact; i.e. $\mathbb{R}\text{-}rank(S_i) \geq 1$ for each i.

Consider an element x in the double coset $C_k = F(k)\backslash\Gamma_k/F(k)$. Since $F(k)$ is Zariski dense in G, we may choose a representative $x \in \Gamma_k$ with $x = vp$ with $v \in U^-(\mathbb{Q})$ and $p \in P(\mathbb{Q})$. Moreover, since the closure of $F(k)$ in the congruence topology on $G(\mathbb{Q})$ is open, we may choose x so that for all primes p dividing the level k, x lies in the open neighbourhood $U^-(k\mathbb{Z}_p)P(k\mathbb{Z}_p)$ of identity in $G(\mathbb{Z}_p)$. Thus the rational matrices v, p have a common denominator, say a; but the elements v, p are integral at all primes p dividing k; in other words, a is coprime to k.

Fix an element $m \in M(a^N\mathbb{Z})$ for some large N. Since the group U^- is normalised by M and $v \in U^-(\mathbb{Q})$ has denominator dividing a, the commutator $[m, v] = mvm^{-1}v^{-1}$ is integral and is divisible by k at all primes p dividing k. Moreover, since $(m, v) \mapsto mvm^{-1}v^{-1}$ is a polynomial in the entries of v and m with integer coefficients, for N large enough, $mvm^{-1}v^{-1}$ is integral at all primes dividing a; in other words, $[m, v] \in V^-(k\mathbb{Z})$, where, we recall, $V^{\pm} = [M, U^{\pm}]$. Similarly, the commutator $[p^{-1}, m] \in (MV)(k\mathbb{Z})$.

We now consider the conjugate mxm^{-1}. We have written $x = vp$ with $v \in U^-(\mathbb{Q})$ and $p \in P(\mathbb{Q})$. Hence

$$mxm^{-1} = mvm^{-1}mpm^{-1} = [m, v]vp[p^{-1}, m] = [m, v]x[p^{-1}, m].$$

Thus, if $m \in M(a^N\mathbb{Z})$, then from the discussion in the preceding paragraph, as an element of the double coset $C_k = F(k)\backslash\Gamma_k/F(k)$, $mxm^{-1} \in F(k)xF(k)$, i.e.

$mxm^{-1} = x$ as double cosets In other words, the group $M(a^N\mathbb{Z})$ acts trivially, under conjugation, on the element $x \in C_k$.

The double coset $x \in C_k$ may be replaced (since $F(k)$ has open closure in the congruence completion of $G(\mathbb{Q})$), by an element $y \in \Gamma_k$ such that for each prime p dividing a, the element $y \in U^-(\mathbb{Z}_p)P(\mathbb{Z}_p)$. In other words, if $y = v'p'$ is written as a product of $v' \in U^-(\mathbb{Q}), p' \in P(\mathbb{Q})$, then $v' \in U^-(\mathbb{Z}_p), p' \in P(\mathbb{Z}_p)$. In other words, the elements v', p' are integral at all primes dividing a. Therefore, the common denominator (say b) of the rational matrices v', p' is co-prime to a. From the conclusion of the preceding paragraph, the group $M(b^N\mathbb{Z})$ acts trivially on the coset representative y of x.

Since $S^*(\mathbb{Q})$ acts on C_k via its image $S(\mathbb{Q})$ in $M(\mathbb{Q})$, we see from the last two paragraphs that, both $S^*(a^N\mathbb{Z})$ and $S^*(b^N(\mathbb{Z})$ act trivially on the double coset $x \in C_k$, hence so does the group generated by these. By Lemma 10, the group generated by these subgroups is $S^*(\mathbb{Z})$, and hence $S^*(\mathbb{Z})$ acts trivially on each coset x in C_k. Hence $S^*(\mathbb{Z})$ acts trivially on C_K and by taking inverse limits, it follows that $S^*(\mathbb{Z})$ acts trivially on the kernel C. But all of $G(\mathbb{Q})$ *acts* on the kernel C, and the infinite group $S^*(\mathbb{Z})$ acts trivially. In view of the simplicity of $G(\mathbb{Q})$ modulo its centre, it follows that $G(\mathbb{Q})$, and hence $\widehat{G}$, act trivially on C: C is central in $\widehat{G}$ and Theorem 4 is proved.

4. The Case Where M is abelian

When M is abelian, the proof of Theorem 4 is more involved. We will use some results from [**V2**]. Since M is abelian and P is a maximal parabolic subgroup, this means that the semisimple part of L has $\mathbb{Q}$-rank zero, and hence $\mathbb{Q}\text{-}rank(G) = 1$. We state them now.

[1] For each prime p, denote by $G_p \subset \widehat{G}$ the subgroup generated by $U^{\pm}(\mathbb{Q}_p)$. The group C is central if and only if, for every pair p, q of distinct primes, the groups G_p, G_q commute. In [**V2**] this is proved only in the case C is a (compact) profinite group (since the main application was to the centrality of the congruence subgroup kernel), but the proof works in general and does not use the compactness of C.

[2] There exists a morphism $\phi : H = SL_2 \to G$ of $\mathbb{Q}$-algebraic groups such that $\phi(U_H^{\pm}) \subset U^{\pm}$. Here $U_H^{\pm}$ is the group of upper (resp lower) triangular unipotent matrices in SL_2. Further, the conjugates $\{{}^s\phi(U_H^+), s \in L(\mathbb{Q})\}$ generate the group U^+.

[3] There exists an infinite subgroup $\Delta \subset M(\mathbb{Z})$ such that for *every* triple a, b, k of mutually coprime integers, the group generated by the collection $\{M(azk + b) : z \in \mathbb{Z}\}$ of subgroups contains this fixed group Δ, and such that the commutator $[\Delta, \phi(SL_2)]$ contains $\phi(SL_2)$.

Assume the above facts, and write $H = SL_2$. For each integer k, write $F_H(k), \Gamma_{H,k}$ for the intersections $F(k) \cap H$, $\Gamma_k \cap H$. Denote by $C_H(k)$ the double coset $F_H(k)\backslash\Gamma_{H,k}/F_H(k)$ and fix an element $x \in C_H(k)$. Then $F_H(k)$ has the same closure in the congruence completion of $H(\mathbb{Q})$ as $\Gamma_{H,k}$. If $x = \begin{pmatrix} a & b \\ c & d \end{pmatrix}$ and $c \neq 0$ (which we may assume after replacing x by a left translation by a suitable element of $F(k)$) we may write $x = vp$ with $v = \begin{pmatrix} 1 & 0 \\ \frac{c}{a} & 1 \end{pmatrix}$ and hence the common denominator of v, p is the integer a. For a suitable power N which depends only on the embedding $\phi : H \to G$, we have that for $m \in M(a^N)$, the commutator $[m, \phi(v)] = m\phi(v)m^{-1}\phi(v)^{-1} \in V^-(k\mathbb{Z}) \subset F(k)$, and the commutator $[(\phi(p))^{-1}, m] \in (MV)(k\mathbb{Z}) \in F(k)$. Consider the conjugate $m\phi(x)m^{-1}$. Writing $x = vp$ we get

$$m\phi(x)m^{-1} = m\phi(v)m^{-1}m\phi(p)m^{-1} =$$

$$= [m, \phi(v)]x[\phi(p)^{-1}, m] \in F(k)\phi(x)F(k).$$

That is, $m\phi(x)m^{-1} = \phi(x)$ as double cosets. Thus the group $M(a)$ fixes the image of the element $x \in C_H(k)$ under ϕ in the double coset $C_k = F(k)\backslash\Gamma_k/F(k)$ where a is the top left entry of the matrix x.

We may replace x by an element $y = x\gamma$ with $\gamma \in F(k)$. We choose $\gamma = \begin{pmatrix} 1 & 0 \\ kz & 1 \end{pmatrix}$ for some integer z. Then $y = \begin{pmatrix} a + bzk & b \\ c + dkz & d \end{pmatrix}$ has top left entry $a + bkz$. By the conclusion of the preceding paragraph, the group $M(a + bkz)$ also fixes the element y viewed as a double coset in C_k. But by construction $x = y$ as double cosets, and hence both the groups $M(a)$ and $M(a+bkz)$ fix x for all $z \in Z$. Thus the group $M_{a,b,k}$ generated by the collection $\{M(a + bkz)\mathbb{Z}) : z \in \mathbb{Z}\}$ fixes the double coset x. By [3] of the listed facts, there is a fixed infinite subgroup $\Delta \subset M_{a,b,k}$ for *every* a, b, k. Hence Δ fixes x for every $x \in C_H(k)$ and hence $C_H(k)$ is fixed by Δ for every k. By taking inverse limits, we see that the image of C_H under the map ϕ is fixed by all of Δ.

However, the image $\phi(C_H)$ of C_H is invariant under the action of $SL_2 = H$. Again by the second part of [3], $\phi(H) \subset [\Delta, \phi(H)]$ acts trivially on $\phi(C_H)$ and hence $\phi(C_H)$ is a central extension of $\phi(SL_2(\mathbb{A}_f))$. Therefore, by fact [1], for each pair of distinct primes p, q the groups $\phi(U_H^+(\mathbb{Q}_p))$ and $\phi(U_H^-(\mathbb{Q}_q))$ commute.

Let $s \in L(\mathbb{Q})$ be arbitrary, and write $s = (s_p) \in L(\mathbb{A}_f)$. Being the linear action, the adjoint action of $L(\mathbb{Q})$ on $U^\pm(\mathbb{A}_f)$ in the topological group $\widehat{G}$ factors through the finite adelic group $L(\mathbb{A}_f)$. Hence, for each p, q, $u \in U_H^+(\mathbb{Q}_p)$ and $v \in U_H^-(\mathbb{Q}_q)$, we have

$$s\phi(u)s^{-1} = s_p\phi(u)s_p^{-1}, \quad s\phi(v)s^{-1} = s_q\phi(v)s_q^{-1}.$$

Furthermore, by weak approximation ([**Gille**]), $L(\mathbb{Q})$ is dense in the product $L(\mathbb{Q}_p) \times L(\mathbb{Q}_q)$. Since $\phi(u)$ and $\phi(v)$ commute by the conclusion of the preceding paragraph, we see (by taking limits of elements in $L(\mathbb{Q}) \subset L(\mathbb{Q}_p) \times L(\mathbb{Q}_q)$) that for every $s_p \in L(\mathbb{Q}_p)$ and every $s_q \in L(\mathbb{Q}_q)$, the elements $s_p\phi(u)s_p^{-1}$ and $s_q\phi(v)s_q^{-1}$ commute for all u, v. But by fact [2], ϕ may be so chosen that the collection $s_p\phi(u)s_p^{-1}$ with $s_p \in L(\mathbb{Q}_p), u \in U_H^+(\mathbb{Q}_p)$ generates all of $U^+(\mathbb{Q}_p)$; similarly for $U^-(\mathbb{Q}_q)$. Hence $U^+(\mathbb{Q}_p)$ commutes with $U^-(\mathbb{Q}_q)$ for each pair p, q of distinct primes. By fact [1], this means that C is central. Thus we have proved Theorem 4 in all cases.

Since we have already shown in the introduction that Theorem 4 implies Theorems 2 and 1, we have also proved Theorem 1 in all cases.

Acknowledgments

I thank the organisers for inviting me to take part in the conference and to contribute an article to the proceedings of the conference.

References

[BMS] H. Bass, J. Milnor, and J.-P. Serre, *Solution of the congruence subgroup problem for* SL_n $(n \geq 3)$ *and* Sp_{2n} $(n \geq 2)$, Inst. Hautes Études Sci. Publ. Math. **33** (1967), 59–137. MR244257

[Benoist-Miquel] Yves Benoist and Sébastien Miquel, *Arithmeticity of discrete subgroups containing horospherical lattices*, Duke Math. J. **169** (2020), no. 8, 1485–1539, DOI 10.1215/00127094-2019-0082. MR4101737

[Benoist-Oh] Yves Benoist and Hee Oh, *Discrete subgroups of* $SL_3(\mathbb{R})$ *generated by triangular matrices*, Int. Math. Res. Not. IMRN **4** (2010), 619–632, DOI 10.1093/imrn/rnp149. MR2595007

[Benoist-Oh2] Yves Benoist and Hee Oh, *Discreteness criterion for subgroups of products of* SL(2), Transform. Groups **15** (2010), no. 3, 503–515, DOI 10.1007/s00031-010-9104-7. MR2718935

[Bour] N. Bourbaki, *Éléments de mathématique. Topologie générale. Chapitres 1 à 4*, Hermann, Paris, 1971. MR0358652

[Gille] Philippe Gille, *Le problème de Kneser-Tits* (French, with French summary), Astérisque **326** (2009), Exp. No. 983, vii, 39–81 (2010). Séminaire Bourbaki. Vol. 2007/2008. MR2605318

[Nori] Madhav V. Nori, *On subgroups of* $GL_n(\mathbf{F}_p)$, Invent. Math. **88** (1987), no. 2, 257–275, DOI 10.1007/BF01388909. MR880952

[Oh1] Hee Oh, *Discrete subgroups generated by lattices in opposite horospherical subgroups*, J. Algebra **203** (1998), no. 2, 621–676, DOI 10.1006/jabr.1997.7188. MR1622748

[Oh2] Hee Oh, *On discrete subgroups containing a lattice in a horospherical subgroup*, Israel J. Math. **110** (1999), 333–340, DOI 10.1007/BF02808188. MR1750430

[Raghunathan1] M. S. Raghunathan, *A note on generators for arithmetic subgroups of algebraic groups*, Pacific J. Math. **152** (1992), no. 2, 365–373. MR1141802

[Tits1] Jacques Tits, *Systèmes générateurs de groupes de congruence* (French, with English summary), C. R. Acad. Sci. Paris Sér. A-B **283** (1976), no. 9, Ai, A693–A695. MR424966

[Vaserstein1] L. N. Vaseršteĭn, *Structure of the classical arithmetic groups of rank greater than* 1 (Russian), Mat. Sb. (N.S.) **91(133)** (1973), 445–470, 472. MR0349864

[V1] T. N. Venkataramana, *On systems of generators of arithmetic subgroups of higher rank groups*, Pacific J. Math. **166** (1994), no. 1, 193–212. MR1306038

[V2] T. N. Venkataramana, *Centrality*, to appear in Michigan math journal, Volume in honour of Gopal Prasad's 75th birthday, (2022). Math arXiv:2108.09006.

[V3] T. N. Venkataramana, *Zariski dense subgroups of arithmetic groups*, J. Algebra **108** (1987), no. 2, 325–339, DOI 10.1016/0021-8693(87)90106-2. MR892908

[W] Boris Weisfeiler, *Strong approximation for Zariski-dense subgroups of semisimple algebraic groups*, Ann. of Math. (2) **120** (1984), no. 2, 271–315, DOI 10.2307/2006943. MR763908

School of Mathematics, Tata Institute of Fundamental Research, Homi Bhabha Road, Bombay - 400 005, India

Email address: venky@math.tifr.res.in

SELECTED PUBLISHED TITLES IN THIS SERIES

783 **Alla Detinko, Michael Kapovich, Alex Kontorovich, Peter Sarnak, and Richard Schwartz, Editors,** Computational Aspects of Discrete Subgroups of Lie Groups, 2023

781 **Emanuel Indrei, Diego Marcon, and Levon Nurbekyan, Editors,** Geometric and Functional Inequalities and Recent Topics in Nonlinear PDEs, 2023

780 **Erik Koelink, Stefan Kolb, Nicolai Reshetikhin, and Bart Vlaar, Editors,** Hypergeometry, Integrability and Lie Theory, 2022

779 **Samuele Anni, Valentijn Karemaker, and Elisa Lorenzo García, Editors,** Arithmetic, Geometry, Cryptography, and Coding Theory 2021, 2022

778 **Carlos Galindo, Alejandro Melle Hernández, Julio José Moyano-Fernández, and Wilson A. Zúñiga-Galindo, Editors,** p-Adic Analysis, Arithmetic and Singularities, 2022

777 **Bang-Yen Chen, Nicholas D. Brubaker, Takashi Sakai, Bogdan D. Suceavă, Makiko Sumi Tanaka, Hiroshi Tamaru, and Mihaela B. Vajiac, Editors,** Differential Geometry and Global Analysis, 2022

776 **Aaron Wootton, S. Allen Broughton, and Jennifer Paulhus, Editors,** Automorphisms of Riemann Surfaces, Subgroups of Mapping Class Groups and Related Topics, 2022

775 **Fernando Galaz-García, Cecilia González-Tokman, and Juan Carlos Pardo Millán, Editors,** Mexican Mathematicians in the World, 2021

774 **Randall J. Swift, Alan Krinik, Jennifer M. Switkes, and Jason H. Park, Editors,** Stochastic Processes and Functional Analysis, 2021

773 **Nicholas R. Baeth, Thiago H. Freitas, Graham J. Leuschke, and Victor H. Jorge Pérez, Editors,** Commutative Algebra, 2021

772 **Anatoly M. Vershik, Victor M. Buchstaber, and Andrey V. Malyutin, Editors,** Topology, Geometry, and Dynamics, 2021

771 **Nicolás Andruskiewitsch, Gongxiang Liu, Susan Montgomery, and Yinhuo Zhang, Editors,** Hopf Algebras, Tensor Categories and Related Topics, 2021

770 **Stéphane Ballet, Gaetan Bisson, and Irene Bouw, Editors,** Arithmetic, Geometry, Cryptography and Coding Theory, 2021

769 **Kiyoshi Igusa, Alex Martsinkovsky, and Gordana Todorov, Editors,** Representations of Algebras, Geometry and Physics, 2021

768 **Dražen Adamović, Andrej Dujella, Antun Milas, and Pavle Pandžić, Editors,** Lie Groups, Number Theory, and Vertex Algebras, 2021

767 **Moshe Jarden and Tony Shaska, Editors,** Abelian Varieties and Number Theory, 2021

766 **Paola Comparin, Eduardo Esteves, Herbert Lange, Sebastián Reyes-Carocca, and Rubí E. Rodríguez, Editors,** Geometry at the Frontier, 2021

765 **Michael Aschbacher,** Quaternion Fusion Packets, 2021

764 **Gabriel Cunningham, Mark Mixer, and Egon Schulte, Editors,** Polytopes and Discrete Geometry, 2021

763 **Tyler J. Jarvis and Nathan Priddis, Editors,** Singularities, Mirror Symmetry, and the Gauged Linear Sigma Model, 2021

762 **Atsushi Ichino and Kartik Prasanna,** Periods of Quaternionic Shimura Varieties. I., 2021

761 **Ibrahim Assem, Christof Geiß, and Sonia Trepode, Editors,** Advances in Representation Theory of Algebras, 2021

760 **Olivier Collin, Stefan Friedl, Cameron Gordon, Stephan Tillmann, and Liam Watson, Editors,** Characters in Low-Dimensional Topology, 2020

For a complete list of titles in this series, visit the
AMS Bookstore at **www.ams.org/bookstore/conmseries/**.